Science in Print

PRINT CULTURE HISTORY
IN MODERN AMERICA

Series Editors

James P. Danky

Christine Pawley

Adam R. Nelson

Science in Print

*Essays on the History of Science
and the Culture of Print*

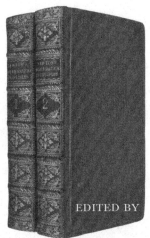

EDITED BY

Rima D. Apple
Gregory J. Downey
Stephen L. Vaughn

THE UNIVERSITY OF WISCONSIN PRESS

Publication of this volume has been made possible, in part,
through generous support from the following departments and units
at the University of Wisconsin–Madison:
the General Library System;
the School of Library and Information Studies;
the School of Journalism and Mass Communication; and
the Robert F. and Jean E. Holtz Center for Science and Technology Studies.

The University of Wisconsin Press
1930 Monroe Street, 3rd Floor
Madison, Wisconsin 53711-2059

uwpress.wisc.edu
3 Henrietta Street
London WC2E 8LU, England
eurospanbookstore.com

Printed in the United States of America

Library of Congress Cataloging-in-Publication Data

Science in print : essays on the history of science and the culture of print /
edited by Rima D. Apple, Gregory J. Downey, and Stephen L. Vaughn.
p.   cm.—(Print culture history in modern America)
Includes bibliographical references and index.
ISBN 978-0-299-28614-9 (pbk.: alk. paper)
ISBN 978-0-299-28613-2 (e-book)
1. Science publishing—History.   2. Scientific literature—History.
3. Communication in science—History.
I. Apple, Rima D. (Rima Dombrow), 1944–
II. Downey, Gregory John.   III. Vaughn, Stephen, 1947–
IV. Series: Print culture history in modern America.
Z286.S4S35   2012
070.5′7—dc23
2011046758

Contents

Foreword

JAMES A. SECORD

What a stupid book!" As the children's author Annie Carey pointed out in *The History of a Book* of 1874, such a dismissal was only possible in an era in which printed materials were common enough for some to be dismissed as trash. Books, journals, preprints, catalogues, and publications of all kinds have long been central elements in the modern world, and specifically to scientific understandings. What is offered in the present book—which, I hardly need add, is far from stupid—is a remarkable conspectus of current approaches to the history of print in relation to science, medicine, and related technical disciplines.

The lively variety that this methodological range encourages is, in large part, because the so-called history of the book, like the history of science and medicine, is (thankfully) not a unified discipline with an agreed approach. Although often portrayed as a single field, the study of print culture is in many ways better seen as an unstable alliance among at least four different approaches. The first grew out of technical bibliographical studies, particularly of the early modern period, and focused on the physical description of printed materials. The second focused on publishers and has tended to use the tools of economic and business history. This has led both to histories of individual firms and to overall surveys of longer periods and different countries—notably James Raven's *The Business of Books* and William St Clair's *The Reading Nation in the Romantic Period*. Third, there has been a variety of literary approaches, most notably centered on understanding forms of address in texts and the expectations brought to texts by potential readers. And finally, cultural historians have looked at the uses of print in society. Much of their writing, as in works such as Janice Radway's *Reading the Romance* and Jonathan Rose's *The Intellectual Life of the British Working Class*, deals with the actual experience of readers. Historians of print do everything from examining minute typographical variations to constructing grand theories about reception.

The great virtue of Robert Darnton's famous "circuit of communication," when first published in 1995, was its openness to all these approaches at once, of allowing a very diverse array of scholars to focus on the book as an object

vii

that passed successively through a range of producers of meaning, from authors to publishers, from publishers to printers, from printers to readers, and from readers back to authors. Itinerant tract sellers and anonymous readers could suddenly stand as legitimate next to the great authors. Not only did it make the simple but crucial point that the meaning of print was the result of many hands, other than those of the author; the circuit also connected a range of hitherto disparate academic enterprises in a nonhierarchical way. There was, in effect, something for everyone to do.

These perspectives have been brought together in a wide variety of works, but the balance between them is likely to remain unstable and subject to fruitful debate. This volume is no exception. Take two of the contributors at opposite ends of the chronological spectrum covered by this volume. In her essay, Meghan Doherty shows how a close attention to seventeenth-century engraving manuals can shed light on contemporary attitudes toward natural philosophy in the Royal Society of London. In doing so, she makes it possible to provide close readings of materials that have previously been described only in general terms. Rima Apple, in contrast, has less to say about the effect of forms of production and the industries of print and mass communication that underwrite her story, but she provides a rich picture of the way printed materials were used to shape public opinion. The difference between the two essays is the result not only of asking different questions but of different ways of thinking about the relation between form and content, between production and use.

If there has been a basis for gathering under a single banner, it almost certainly lies in a focus on the material basis of communication. Whether seen as objects of bibliographical enquiry, manufactured objects, or cultural goods circulated in society, books are constrained by their materiality. Books, pamphlets, handbills, journal issues, and all other printed documents are things; their making and fortunes can be traced just like any other objects (to use a language that has become current since writing about print became widespread). It matters that they have bindings, typefaces, paper, and prices; it matters that they can travel in some places easily and other places only with difficulty or not at all.

Not only does this attention to material form bring different approaches to print into a closer alliance, but it also suggests connections to recent work in history of science and medicine. For the interest currently enjoyed by studies of the material culture of print is part of a wider shift toward a history in which practices, rather than abstract concepts, are given pride of place. It is not surprising, then, that historians of science and medicine have turned away from the rarefied world of the history of ideas to focus on what scientists do, what Bruno Latour called "science in action," just as studies of print have achieved ascendency. Experiment, fieldwork, collecting, lecturing, and museum display have become central at the same time as writing, printing, publishing, and reading. From Adrian Johns's *The Nature of the Book* to Sally Kohlstedt's *Teaching Children Science*, historians of science have played a strikingly large role in studies of print, much greater than their numbers would imply.

Just as in Darnton's communication circuit, historians of science and medicine have brought into play a much larger range of producers than the handful of heroic authors who dominated traditional accounts. Those whom Steven Shapin called "the invisible technicians" of science have proved not so invisible after all, as laboratory workers, map colorers, working-class botanical collectors, and others have been restored to their position in the cycle of scientific labor. In medicine, too, it has long been established that a history based on the "great doctors" needs to be replaced by a much wider range of participants.

A history of science and medicine from below has led to a burgeoning field of studies of the uses of specialized knowledge in everyday life. Important foundations for this were laid, as in so much else in current practice, by studies of the gendered nature of intellectual labor. Work by Anne Shteir, Ludmilla Jordanova, Dorinda Outram, and many others opened up the possibility of understanding the significance of mundane practices in the making of knowledge. By the early 1990s, and particularly after a *History of Science* special issue, with its much-cited review essay by Roger Cooter and Stephen Pumfrey, the history of popular science has emerged as an exciting and innovative field. We have a much better idea than before about the role of science journalists, the making of educational documentaries, and the creation of mass audiences for science.

Much of this work has involved approaches drawn from the history of print culture, so much so that for the nineteenth and twentieth centuries the history of what is often termed "popular" science seems identical with the history of science in print. As Lynn Nyhart points out, historians drawing on these tools have focused almost entirely on the nonspecialist end of the market. We know more about the Religious Tract Society as a publisher than the Royal Society, more about science in general periodicals such as *Blackwood's* and the *Quarterly Review* than in most of the specialist ones, more about the publishing of the anonymous evolutionary tract *Vestiges of the Natural History of Creation* (1844) than *On the Origin of Species* (1859). The situation is especially striking in Britain but is also present in studies of other European countries and the United States.

Having contributed to this literature, I would certainly not want to undervalue it, both in its own terms and in terms of the wider lessons that have been learned. There are, however, pitfalls, not all of which are easy to avoid. Not least, hiving off studies of "popular science" into a separate arena could be taken to imply that studies of its supposed opposite, "elite" or "specialized" knowledge, can carry on unchanged. To say that a work transforms our understanding of popular science could be taken as implying that it does not change at all our understanding of science proper. Approaches drawn from print culture, it could then be assumed, are only relevant when large audiences, without direct engagement in the practice of science, are involved. At its most damaging, the implication is that communication practices within science are effectively transparent. Such a view tends, in an entirely unanticipated way, to reinforce the much criticized "diffusionist" model of scientific communication, which treats knowledge among so-called expert and popular audiences asymmetrically.

It is certainly time, then, that the circuit of communication and the cycle of knowledge-making intersected more extensively. For this to happen, historians will want to follow the essays in this collection in attending further to questions of the mid-level analytical tools for understanding print.

One of these tools, surely of crucial importance in going beyond the simple dichotomy of "popular science" and "specialist science," is genre. Genre is widely dealt with in literary scholarship but largely ignored in history of science. An understanding of genre makes it possible to identify the range of common expectations that readers, authors, and publishers bring to the works they use. There are many examples here, whether these be evolutionary treatises, printers' manuals, advertisements, or children's books. Genre, it should be stressed, does not offer a stable set of pigeonholing categories; like class and gender, it is a mode of analysis. No work ever belongs to a single genre. This is abundantly clear in the present volume, nowhere more so than in Kate McDowell's wonderful demonstration of the ways even such dry tomes as Charles Darwin's *Variation of Animals and Plants under Domestication* (1868) could appear on a list of evolutionary books recommended for children. Similarly, a single book could change its status, as Cheryl Knott suggests in her account of Stewart Udall's *The Quiet Crisis*. The first edition, published in 1963, when Udall was secretary of the interior, was a paperback bestseller and widely read as an influential polemic in the nascent environmental movement. By the time it was reissued with extra chapters during the Reagan era, it was a minor contribution to the historical literature, with limited circulation and little influence. It was, effectively, two different books.

Issues of genre are as apparent in the physical qualities of books as they are in what readers make of them. This is especially important in understanding elementary textbooks, which have traditionally proved recalcitrant to historical analysis. Sally Kohlstedt's work on early-twentieth-century nature study shows, as does her recent book, how an understanding of text as object and nature as object can come together. It is also evident in Robin Rider's remarkable account of textbooks used for teaching mathematics in the early American republic. Through an appreciation of seemingly trivial, and even incidental, differences in typesetting, Rider is able to point to important differences in the way the subject is presented and the nature of the intended audience. Genre here becomes not a simple means of classification but a subtle way to chart ongoing debates about how mathematics should be taught, and what it was for. Although concerned primarily with elementary works, this approach surely deserves to be applied to mathematical journals and monographs.

In fact, one of the most surprising features of the history of science and medicine of the past few decades is that we still have such rudimentary overviews of the generic forms of scientific publication. We have no histories of the development of the laboratory manual, the field guide, the history of technical publishing societies, and the general reflective survey. Most surprisingly, there is still no good history of the scientific journal and the rise of the scientific article as a dominant form of publication. At least it is becoming clear that the scientific article did not arise fully formed with the first number of the *Philosophical Transactions* of the Royal Society but is primarily a nineteenth-century invention, whose dominance is now increasingly challenged by electronic media.

Nyhart's chapter is a particularly significant effort to analyze a key form of scientific publication: the expedition report. Building on the reputation of such monumental classics as Alexander von Humboldt and Amié Bonpland's thirty-volume *Le voyage aux régions equinoxiales du Nouveau Continent, fait en 1799– 1804*, and the even more massive, government-sponsored *Description de l'Egypte*, the "scientific results" of a wide range of expeditions were published in the nineteenth and early twentieth century. Using published *Results* of the Plankton Expedition, an enterprise initially organized by the Prussian state, Nyhart deftly shows how the part-publication of these works, over many years, shaped their distribution, observational and reporting techniques, and use within the relevant specialist communities. Serial publication also provided a focus for stability within a research group, for training new recruits, and for continuity within the research program.

Maintaining publication of serial works over many decades, through war and changing personnel and organizational upheaval, posed serious challenges to the international networks of exchange for science. How were such books to be obtained, catalogued, and categorized? Where were they to be kept? That these questions are only beginning to be tackled is one indication of the Cinderella status the history of libraries occupies within the study of science and medicine. Unlike museums, which have been at the center of attention in these fields, libraries are too often taken for granted. Systems for classifying, organizing, and preserving knowledge are only beginning to be studied, and this is another area in which the present volume suggests avenues for further work.

Bertrum H. MacDonald's chapter shows how the Smithsonian Institution in Washington served as a central clearing house for knowledge, particularly between naturalists in different countries. Agricultural scientists such as William Saunders in Ontario and collectors Roderick MacFarlane of Hudson's Bay could keep up with recent periodicals through the facilities provided by Spencer Baird and the other secretaries of the Smithsonian. Such men were effectively

agents of global scientific diplomacy, part of the reason the half-century before
the First World War was a golden age for global communication within the
sciences. Translation into the main world scientific languages—German,
French, and English—was virtually assured for any contribution of significance,
and enhanced postal systems gave sending publications across national boundaries an unprecedented degree of security. This growing sense of a global literature led to attempts to rationalize and control it. As Jennifer J. Connor suggests,
George M. Gould attempted to accomplish these goals for medical journalism,
to the extent of attempting to standardize the very language of medical communication. The centerpiece of his proposals was the creation of a centralized
exchange of medical literature among libraries.

Debate about the proper relation between knowledge and commerce
continues, and librarians and historians have even more pressing reasons to
talk. The issues involve not only our view of history but the very existence of the
materials we use for writing it. Severe cuts in public spending have led libraries
to take long, hard looks at their existing collections. Scientific and medical
monographs and journal runs are, like most materials from the nineteenth and
twentieth centuries, consulted infrequently. A typical pattern of use will include
a vigorous period of intense consultation by a single scholar, preceded and
followed by quiet decades of undisturbed life on the shelf. How many people at
the University of Minnesota looked at the fine collection of nature study textbooks before Sally Gregory Kohlstedt asked to see them? How often has any-
one had cause to consult the German-language expedition reports so effectively
used by Lynn Nyhart? Today, the accessibility and even the very existence of
such collections are in many places under threat.

One real danger is that large numbers of these works will simply be thrown
away, caught between specialist libraries that no longer need them for current
scientific work and humanities collections that fear them as space-hungry
invaders. In Britain, the UK Research Reserve scheme aims, in part, to avoid
the worst consequences of a lack of coordination in disposals by ensuring that
at least three copies of every work are kept in public collections. At the same
time, however, this potentially opens the door to large-scale dismantling of
historic collections throughout the country, with the excuse that the three-copy
rule will make certain that works remain "accessible" through interlibrary loans
and photocopies of individual articles. As any historian knows, the application
of this policy on a national basis could destroy the possibility for serious academic
work, particularly in periodicals, where consultation of a limited number of
volumes of a series on an occasional basis is no replacement for detailed work
with an entire run.

A more promising solution is through the provision of full runs of journals
in electronic form, particularly through JSTOR and other carefully digitized
projects. (Google Books, with its complete lack of bibliographical control, is not

a serious scholarly resource, however useful it may be.) Although digitization can help solve the problem of access to complete runs, it also carries obvious dangers of its own. At present, most such projects do not include advertisements; their reproduction of illustrations is entirely inadequate; and they do not carry anything like the same amount of bibliographical information. We need, I suspect, to rethink our definition of "special collections," which too often are focused on curation of donated collections rather than materials gathered by decades of careful work by librarians and academics working closely together.

As a postgraduate student at Princeton many years ago, I was lucky enough to learn about new approaches to the history of science at the same time that Robert Darnton was conducting his pioneering research there into the publishing history of the *Encyclopédie*. My own approach to these subjects was deeply shaped by an early exposure to great collections such as the Huntington Library, the Pierpont Morgan Library, the New York Public Library, and the libraries of the University of Wisconsin. I remember sitting in the Geological Society of London in Burlington House, surrounded by shelves of geological proceedings and transactions, and realizing that these forgotten works—and not only a handful of treasured theoretical treatises—would be critical to creating a history based on practice. It would certainly be ironic if great library collections were dispersed or destroyed, just as we realized that in them rests the key to a new kind of history.

Science in Print

Introduction

STEPHEN L. VAUGHN, RIMA D. APPLE, and
GREGORY J. DOWNEY

As the crucial means of recording, distributing, and consuming knowledge for centuries, the practices and products of print culture—books and journals, pamphlets and posters, newspapers and magazines—have been essential to virtually every human endeavor. This fact is especially striking in the arenas of science, technology, engineering, and medicine (STEM), four interrelated fields of systematic knowledge production tied together by their dependence on an accurate and accessible cultural record of empirical data, imaginative theory, and political debate. Whether on paper, as grade-school science textbooks and specialist engineering journals, or, as is increasingly the case, online, as databases of medical advice and popular explanations of scientific principles, the artifacts of print culture have always been the primary means for both making scientific claims and bringing those claims to the public.

Growing levels of global literacy have allowed scientific ideas to flourish as never before, and increasing connections within the global economy have stimulated demands for scientific ideas at an unprecedented pace. The efficient and ethical creation and circulation of ideas in the STEM fields is central not only to the global economic engine but also to those attempting to address profound global challenges such as the consequences of climate change, the development of sustainable energy systems, the provision of clean water, and improvements in public health. Even beyond these important goals, the STEM fields, and the print culture they continue to foster, represent our noble aspirations to explore and understand our environment, our physical universe, and ourselves.

For all of these reasons, in fall 2008 the Center for the History of Print and Digital Culture (CHPDC) at the University of Wisconsin–Madison sponsored an international conference on "The Culture of Print in Science, Technology, Engineering, and Medicine."[1] Organized primarily through the UW–Madison School of Library and Information Studies, under the leadership of CHPDC director Christine Pawley, the meeting was cosponsored by the Wisconsin Historical Society, the University of Wisconsin Libraries, the Department of English, the Department of Curriculum and Instruction, and the School of Journalism and Mass Communication. As this list suggests, the historical study of print and digital culture at Wisconsin is, quite intentionally, an interdisciplinary pursuit. But what made this conference cross even more boundaries was the cosponsorship and participation it received from the STEM fields on campus, particularly through the Holtz Center for Science and Technology Studies, the Department of the History of Science, and the Department of Medical History and Bioethics. This combined focus on the tools, techniques, and meanings developed through centuries of print culture in scientific practice drew dozens of conference attendees from all over the world. Coming from engineering programs and medical faculties as well as library schools and humanities departments, participants spent three days together in sessions with such titles as "Homing in on Engineering Education," "Popularizing Science in Print," "Hybridity: Science and Literature," and "New Technologies and Graven Images," to name a few.

Nine essays from the conference were chosen for inclusion in this volume. Because they vary widely in terms of their topics, time periods, and points of attachment between print culture and scientific knowledge-making, they each help to illuminate the broad history of what librarian and print-culture historian Robert Darnton calls "circuits of communication," incorporating not only different information modes and media technologies, but different types of public events and social milieus as well.[2] The three sections of this volume correspond roughly to three key developments in the circuit of communication: production, distribution, and reception. This simple categorization, however, should not imply a simple linear process of development. As the essays here demonstrate, scholars of both print culture and scientific knowledge production must attempt to connect the fundamental social processes of communication — exchanging information, enabling action at a distance, and participating in a shared symbolic culture — to several centuries of changing social norms, political-economic environments, and information infrastructures.[3] Our volume is meant to demonstrate the myriad ways the history of print culture enriches the history of science and how the history of science informs the history of print culture. Thus, each essay stands both as a case study and as an indication of further lines of inquiry.

The first section of this volume, "Natural Philosophy and Mathematics in Print," focuses primarily on the production of scientific literature by exploring the way the tools and practices of print culture have structured and influenced the development of scientific, technological, and medical knowledge. While it is customary within print culture to explore the history of scientific and technological change in paper-making, or book cataloging, or newspaper production, or photographic duplication, it is less common to consider the effect new technologies of printed text and image might have on scientific thought and practice itself. Historical developments in print culture, from the engravings of the seventeenth century to the typography of the nineteenth century to the photographic illustrations of the twentieth century, have helped foster a new understanding of scientific developments within and among scientific and lay communities.

As Meghan Doherty shows in her chapter "Creating Standards of Accuracy: Faithorne's *The Art of Graveing* and the Royal Society," the printed word and the creation of visual representations have been closely related. The guidelines for preparing exact engravings of the natural world during the mid-seventeenth century resulted from the cooperation between artisans and members of the Royal Society of London. This collaboration put great emphasis on developing precisely the right language to create the "grammar and syntax of accuracy." Without this lexicon, progress in the art of engraving would have been hindered. Printed texts that dealt with the process of engraving made it possible for artisans to craft more realistic images of nature. But they also reinforced claims about the reproducibility of knowledge about that same nature.

Considering the visual impact of print culture does not always involve studying illustrations, however. In her chapter, "'Perspicuity and Neatness of Expression': Algebra Textbooks in the Early American Republic," Robin Rider shows that even the printed page had its own visual culture. Rider maintains that during the late eighteenth and early nineteenth centuries, the way printed words appeared on the page in mathematical books, such as algebra texts, was central to judgments about the importance of those works. Readers and scholars often drew conclusions about the quality of mathematical texts, in part based on their typography. "That which was clear, perspicuous, and neat was held in high esteem," she says. Superior typography underscored and reinforced ideas while "indifferent typography" obscured thought and confused readers. In the words of the political scientist and student of information design Edward Tufte, if the typeface was executed well, it presented "enormous advantages of seeing." Typography, Rider explains, "served to make important mathematical points in and of itself." How algebraic expressions lay, black on white, on the printed page helped readers better understand mathematics. Rider's essay clarifies the sometimes complex interrelationship between the page layouts of textbooks and the comprehension of complicated ideas. Her work also suggests that a judgment about the value of a scientific idea often cannot be separated from a judgment about the visual presentation of that idea.

Both these case studies of the "production" of the print culture of STEM suggest new areas for future research. Doherty's essay points to the critical role of technicians in the creation of scientific ideas; Rider's reminds us that we must consider the practical and the intellectual consequences of layout when studying scientific texts. In so doing, these chapters broaden the scope of the history of print culture and the history of science.

The second area our volume explores is the role print culture has played in the dissemination of knowledge within and without scientific communities. Scientists and researchers have long utilized the power of print culture to communicate with their colleagues, alert others to their discoveries, and foster the exchange of ideas over time and space. The techniques and possibilities of print technology—especially in its translation to the digital realm—continue to help illuminate and clarify scientific theories, findings, and debates. In the section of our volume titled "The Circulation of Scientific Knowledge in Print," three of our essayists document the print culture's critical work in spreading scientific ideas across borders both geographical and conceptual and establishing and maintaining broad scientific networks.

In 1889, the German steamer *National* carried six scientists and an artist who collected plankton from the Atlantic Ocean in an effort to determine the distribution of this important food source for marine life. In her chapter "Voyaging and the Scientific Expedition Report, 1800–1940," Lynn K. Nyhart argues that one of the most important aspects of the German Plankton Expedition was the commitment on the part of the scientists involved to publish their findings. These reports, produced by scholars from different geographical regions, analyzed a variety of specimens, and those analyses were then sent to a project editor to be published. The widespread distribution of these studies represented a central characteristic of large-scale science during the late nineteenth century. Unlike many nineteenth-century scientific voyages, in which specimens were collected but then languished unanalyzed in boxes for years thereafter, the Plankton Expedition resulted in more than sixty separate reports published under the title *Results of the Plankton Expedition of the Humboldt Foundation.* These reports, Nyhart contends, are important to understanding the interconnection between the history of science and print culture. They differ from much recent writing that has focused on popular science, and they are of a different order, for that matter, from the knowledge transmitted through encyclopedias, dictionaries, textbooks, or the more analytical work that is published in peer-reviewed journals. Nevertheless, the printed studies that appeared in *Results,* all of which except two appeared between 1892 and 1913, shaped "the ongoing intellectual substance of zoological science" and "became an integral part of an entire social system of scientific work."

In "Crossing Borders: The Smithsonian Institution and the Diffusion of Scientific Information between the United States and Canada in the Nineteenth Century," Bertrum H. MacDonald explores more explicitly the fact that print has been essential to the transmission of scientific knowledge across national boundaries. MacDonald focuses on the Smithsonian Institution in Washington, D.C., established in 1846 to expand knowledge and increase its diffusion around the world. Thirty years after James Smithson's initial bequest, the Institution had become the world's foremost international distributor of scientific publications. Although historians have written about the Smithsonian's official international exchange program, MacDonald chronicles the extensive stream of information that flowed through its less formal networks of distribution. Using the Smithsonian's archives, he examines the correspondence of its first secretaries, Joseph Henry and especially Spencer Baird, to show how letters and publications disseminated knowledge between the United States and Canada and stimulated scientific research in both countries. During Baird's thirty-five-year tenure, "tens of thousands of letters flowed in and out" of Washington to a vast network of individuals in North America and elsewhere. These people looked to Baird for support and counsel as well as to be "a link to the outside world." Baird was one of the foremost collectors of natural history during the nineteenth century, and he used his correspondence with Canadian scientists to expand the Smithsonian's collection of specimens. In return, he provided amateur and professional scientists in Canada with a large complementary exchange of published and unpublished scientific materials. As a result of these kinds of exchanges across borders, by the end of the nineteenth century, "science had reached a new level of activity."

Print was a vehicle not only for the exchange of scientific data and theory, though. Scientific dispute and debate are also carried across borders through the culture of print, as Jennifer J. Connor explains in her chapter, "Writing Medicine: George M. Gould and Medical Print Culture in Progressive America." In the history of print culture, medical journals and other medical publications such as books, as well as the role of medical editors, authors, and their professionalization, have received relatively little attention from historians. The ethics of medical publishing was of deep concern to such men as George M. Gould, an internationally respected and outspoken American medical editor. Between 1898 and 1906, Gould edited three weekly medical journals, and he believed that the control of medical journals should reside firmly in the hands of physicians. It was important, he held, to know who published medical information and to be certain that the editors and authors were of high character. Literature produced by self-promoters and others of dubious character should be discredited. He was appalled that "whole articles and books and 'systems'" existed in which "not a page . . . was written, and often not read, by the men credited with the authorship." He wanted to eradicate such unscrupulous practices. He worked to ensure that only high quality medical information would be disseminated. In 1898, he set up the Association of Medical Librarians in an effort to increase the circulation of good medical literature, and he criticized medical publishers who declined to give their journals and books to libraries. The American Medical Association, which supported the creation of medical libraries in all U.S. communities, soon endorsed Gould's plan and donated its journal free to members of the Association of Medical Librarians. Gould stood as "the lone voice of the sentinel," calling attention to the ethics of medical publishing, Connor writes, and "thus serves historically as a powerful lens through which to view the medical profession's concern about ownership and distribution of medical knowledge" during the late nineteenth century. He "articulated concerns virtually identical to those of editors around the globe today," Connor contends, but did not have the advantage of our time's sophisticated international communication infrastructure.

Taken together, the essays by Nyhart, MacDonald, and Connor move us well beyond the limited category of "circulation." Concentrating on both the processes and the pressures of publishing, they illustrate ways scientific communities are created, maintained, and extended. Who provides access to what knowledge and under what conditions reveals much about the power of scientific networks, as well as the politics underlying their establishment.

Finally, the third area our volume explores reminds us that the influence of print culture in the scientific and technical fields has not been limited to professional communities. Print culture has always been an essential component of popular education as well, bringing the claims of scientists, technologists, engineers, and medical practitioners to a variety of audiences. Print culture, even in its transition to new digital forms, remains the site where political and philosophical questions about science, technology, and medicine are played out in the public sphere. The final section of our volume, "Science Education and Health Activism in Print," examines such debates.

In her chapter "Evolution in Children's Science Books, 1882–1922," Kate McDowell shows that scant attention has been paid to how children's science books presented evolution during the late nineteenth and early twentieth centuries. McDowell analyzes such works that were recommended by librarians from 1882 to just before the Scopes trial in 1925. "While there were many science books published for children in the United States," she says, "relatively few of them described or discussed evolution, and fewer still explained human evolution." In fact, of the more than two hundred books recommended by librarians, only nineteen mentioned evolution. McDowell speculates on why so few children's books took up this topic and concludes that the answer was not so much political or religious in nature as it was a practical problem. The immense scale of geological time involved in evolution was difficult for young minds to comprehend, and conveying it simply was equally difficult.

McDowell's chapter is interesting not only for what it says about the state of information young people had on evolution before the Scopes trial but also for the method she employs to recover this corner of the past. She uses digitized texts available online via such services as the Internet Archives, Project Gutenberg, and Google Books. The digitization of written texts offers a powerful new way of mining the past and has great potential for transforming what we can learn from history. These massive databases make it possible to search randomly books and articles for names and terms and to find them within a matter of seconds. In the pre-digital era, finding such historical information in printed texts or on microfilm often required months, if not years, of painstaking searching. The proverbial needle in the haystack, which in the past was too often lost, now can frequently be located with ease. McDowell uses digital searches to locate such terms as "Darwin," "evolution," "natural selection," and more in these children's science books. It is important to note, though, that she also combined these digital searches with "in-depth reading and examination of the texts."

When it came to teaching about nature during the late nineteenth and early twentieth centuries, conventional wisdom held that the best way to inform students was not by having them read books but by giving them hands-on experience working in nature itself. In "'Through Books to Nature': Texts and Objects in Nature Study Curricula," Sally Gregory Kohlstedt writes about the dilemma facing progressive educators who wanted to incorporate printed materials into the study of nature. She examines the work of three authors who wrote guidebooks for teachers: Charles A. Scott, Clifton F. Hodge, and Anna Botsford Comstock. Between 1900 and 1920, instructors who taught students about nature often read and assigned the works of these three authors. "The paradox of emphasizing that nature was best studied through objects while producing books relating to nature to encourage that practice was not lost on nature-study advocates," says Kohlstedt. Each author approached the topic differently, but their writings stimulated teachers and gave them innovative ideas. Scott, Hodge, and Comstock treated teachers as "peers and collaborators" in advancing nature-study programs. They made liberal use of illustrations, photographs, and other visual materials in their attempts to inspire readers. Comstock, for example, sought to place children "in sympathy" with the natural world and to kindle their imaginations. She urged them to record their observations of nature. Comstock, like Scott and Hodge, sought to open the minds of students to "the ordinary life around them." For these three authors, Kohlstedt writes, "books were complementary to direct contact with the natural world, sometimes a stimulus to engage in such studies, sometimes a way to discover how personal knowledge fit with more established wisdom, and sometimes a contribution to enhancing fundamental understanding of a phenomenon."

Moving into the twentieth century, Rima D. Apple discusses the important place print culture has occupied during the past century in informing, and sometimes confusing, the public about the science of nutrition. In her chapter "Basic Seven, Basic Four, Mary Mutton, and a Pyramid: The Ideology of Meat in Print Culture," Apple explains how print has been used to promote an "ideology of meat." Meat has been at once "highly controversial" and "very popular," Apple writes. "In the drive to educate Americans, primarily female consumers, about the nutritional value of meat in a healthful diet, we see the power of print culture, of text and graphics, to shape the dimensions of the discussion." When one looks at efforts to educate women about food in government publications, books, periodicals, and the ads that accompany them, it's clear they "all employed the language of science—whether it came from the laboratory, the field, or the copywriter's pen." And yet, as the world of print proclaimed meat's importance, Apple observes, it did so "without any consensus on its definition" or what its place in a healthy lifestyle should be. During the late twentieth and early twenty-first centuries, medical science has pointed to the dangers of eating excessive amounts of red meat, especially meat with a high fatty content. But even as the evidence showing the connection between certain types of meat and coronary artery disease mounted, much information continued to promote an image of meat as not merely "a good source of protein, but as *the* definition of food," Apple maintains. The print culture of such things as the Basic Four, the Food Pyramid, and Molly Mutton plainly strengthened the ideology of meat, at a time when much other printed material "bombarded the American public with scientific arguments against meat." Any attempt to promote a healthier diet in the future, Apple contends, must develop a "new schema for nutrition education, textual and graphic, that does not overtly and inadvertently reinforce the ideology of meat."

Our final chapter brings our path through the intertwined history of print culture and the scientific fields up to the latter half of the twentieth century. In this period, when motion pictures, television programs, and computers grew in popularity, Cheryl Knott reminds us that, according to Stewart Udall, books remained a powerful force for the "reorientation of human thought." Indeed, she argues, books have not just survived but have often flourished and have been catalysts in launching political and social movements. Books, more than cinema or TV, give authors control over how their ideas will be expressed, although, as Knott acknowledges, they do not give them "control over how the information is received and used." Fortunately, books provide historians with a level of information not readily available elsewhere. They can be important sources to document the origins of ideas and arguments that can only be presented superficially in interviews and speeches. In her chapter "What Two Books Can (and Cannot) Do: Stewart Udall's *The Quiet Crisis* and Its Twenty-Fifth Anniversary Edition," Knott examines the context and impact of a book that warned of the deterioration of America's natural resources as it reminded readers of the history of the nation's conservation movement.

First published in 1963, *The Quiet Crisis* was expanded by the author and reissued in a twenty-fifth anniversary edition in 1988. When it first appeared, Udall was secretary of the interior under President John F. Kennedy, a post he would retain through the Lyndon B. Johnson administration. Udall had been influenced by an earlier book, Knott maintains, Rachel Carson's *Silent Spring* (1962). Using Udall's papers at the University of Arizona, Knott argues that the first edition of Udall's book was much more influential than the later expanded edition. She raises an interesting issue regarding the authorship of best-selling commercial works. The writing of both Carson's book and Udall's "involved a network of people." In Udall's case, several people provided information, drafted sections, and offered other advice. For the anniversary edition, Udall and his assistants added an additional eight chapters to the original text, including a critique of the Reagan administration's neglect of environmental issues. Despite these additions, the expanded volume received only a fraction of the attention given to the original. According to Knott, the difference in the reception given to the two editions can be explained partly by the fact that Udall was no longer in the president's cabinet but more by the great growth of environmental science and politics between 1963 and 1988. The expansion of the environmental movement owed much to a flood of books and articles written after the first appearance of Carson's *Silent Spring* and Udall's *The Quiet Crisis*.

All four of these chapters highlight the importance of evaluating the broader context in interpreting print culture in science education, whether in schools or in the wider public. As James Secord notes in his foreword to this book, the public dissemination of scientific knowledge has been a popular theme in the history of the print culture of STEM. Still, the chapters in this section suggest that it remains a rich field with additional, important, and potentially fruitful areas that need to be studied. Both McDowell and Kohlstedt interrogated children's literature, demonstrating that texts cannot be studied outside of the culture that forms them. In McDowell, we see the difficulties authors had in conveying a scientifically appropriate story within the confines of contemporary pedagogical philosophy. In Kohlstedt, we see how contemporary pedagogical philosophy helped to create textbooks even in an area of study that scorned them. Apple and Knott place their texts within the changing social and political values of the period, values that can at one time give primacy to certain ideas—such as the importance of meat as the source of protein and the newly emerging awareness of environmental concerns—while at other times minimize them.

Whether documenting the significance of typographical layout in the shaping of scientific ideas, the critical role of institutions and individuals in building scientific communities, or the pivotal function of a text in shaping a curriculum, the nine essays in this book represent only several of the emerging paths intertwining the history of print culture and the STEM fields. Even this small sample demonstrates the potential of such interdisciplinary scholarship. As Florence C. Hsia's "Note on Sources" indicates at the close of our volume, combining a rich analysis of print culture — in all its economic, aesthetic, professional, educational, material, and social aspects — with the history of science, technology, and medicine reveals new and thoughtful perspectives on the creation and dissemination of scientific knowledge. In analyzing the details of production under various political-economic regimes, the mechanisms of distribution through nascent knowledge networks, and the contradictions of circulation through a diversity of audiences, this collection of essays brings new insights to the study of science, its actors, and its cultural context.

Notes

1. "School of Library & Information Studies, University of Wisconsin–Madison," accessed March 30, 2012, http://www.slis.wisc.edu/chpchome.htm.
2. See for example Robert Darnton, "An Early Information Society: News and Media in Eighteenth-Century Paris," *American Historical Review* (February 2000): 1–35.
3. See for example Gregory J. Downey, *Technology and Communication in American History* (Washington, D.C.: American Historical Association, 2011).

Natural Philosophy and
Mathematics in Print

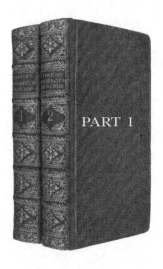

Creating Standards of Accuracy

Faithorne's The Art of Graveing *and the Royal Society*

MEGHAN DOHERTY

> For my part, I prefer (to guard the Dead)
> A Copper Plate, before a sheet of Lead.
> So long as Brasse, so long as Books endure,
> So long as neat-wrought Pieces, Thour't secure,
> A *Faithorne sculpsit* is a Charm can save
> From dull Oblivion, and a gaping Grave.
>
> Thomas Flatman, "To My Ingenious Friend Mr. Faithorne"

This essay argues that standards of accuracy developed over the course of the seventeenth century and came to govern the printed representation of the natural world as a consequence of being deployed in the production of images used for natural history and natural philosophy texts; furthermore, it shows how such standards were developed through the collaboration of artisans and members of the Royal Society of London. As a result of these standards, engraving was treated as a transparent medium; that is, by describing and then conforming to a set of commonly accepted standards, engravers projected a direct relationship between the image and the natural world.[1] I do not mean that these images were "truthful" or "objective" by some transhistorical measure.[2] Instead, let me suggest, that these images facilitated the production and circulation of knowledge within a group, the Royal Society, and further, these images

15

were contributors to the construction of what Steven Shapin has called the
"social history of truth."[3] This essay, then, contends that the engravings produced for the Royal Society in the seventeenth century played a critical role in
the circulation of knowledge because they participated in the gentlemanly practices of witnessing that were essential to the Royal Society's attempts to validate
the knowledge its members produced.[4] Members of the Royal Society included
intaglio printed images in their books as stand-ins for the objects depicted despite
the great distance between each object and its representation. These images
need to be situated in a social context because, like the "fact," they represented
mediated knowledge.[5] Both artisans and members of the Royal Society were
concerned with how prints were made and how they functioned as if they trans-
parently represented the natural world. The "as if " of this proposition is key: it
is essential to highlight the strangeness of the possibility that a small, flat, color-
less, line image on paper with no simulation of touch, sound, or smell could
be taken to represent the full and actual presence of an object. To show how
printed images became a necessary part of the communication of knowledge, I
aim to unpack how they were produced and how the process of engraving was
written about in the middle of the seventeenth century.[6]

I have chosen the Royal Society as the locus for this study of the development of standards because of the important role it played in promoting and
facilitating the increasingly widespread interest in natural knowledge in the
seventeenth century, and, moreover, for its avowed commitment to accuracy
and exactness. Founded in 1660 for the promotion of experimental learning,
the Royal Society worked to improve the state of knowledge about the natural
world through experiment and observation.[7] The Fellows of the Royal Society
broke with past reliance on ancient opinions regarding the workings of Nature
and instead relied on their own experiences.[8] As Thomas Sprat wrote in his
History of the Royal Society (1667): "I shall lay it down, as their *Fundamental Law*,
that whenever they could possibly get to *handle* the subject, the *Experiment* was
still perform'd by some of the *Members* themselves. The want of this *exactness*, has
very much diminish'd the credit of former *Naturalists*."[9] It should be noted at
this point that the word "accuracy" was coming into the English language at
exactly the same moment as the Royal Society was founded and that early uses
of the word occur in texts by its Fellows.[10] The standards of accuracy that
governed the production of engravings for texts published by Fellows developed
within this context of experiment, which valued knowledge gained through
first-hand experience. These standards helped to ease the transmission of
knowledge as it passed from one individual's observation into a print that could
circulate among a larger audience.[11]

While Steven Shapin's work on the social history of truth focuses on gentlemanly culture in mid-seventeenth-century London, I argue that the standards
of accuracy that helped intaglio printed images to "pass" for transparent

representations of nature developed out of the writings of an artisan: William Faithorne (circa 1620–1691). Faithorne was not working in a completely separate sphere from the gentlemanly culture of the Royal Society; on the contrary, his work was intricately twined with that of the Fellows. Faithorne is known to have drawn portraits from the life of nine Fellows of the Royal Society.[12] He is one of the three artists who signed plates included in John Ray's posthumous publication of Francis Willughby's *Ornithologiæ* (London, 1676).[13] In addition, Faithorne engraved the only signed plate published in the *Philosophical Transactions of the Royal Society* during Henry Oldenburg's tenure as editor of the journal, 1665–77.[14] Faithorne was also sought after later in his career to make drawings that other artists then engraved for the *Philosophical Transactions*.[15] These brief examples show that Faithorne was deeply involved in the visual communication of knowledge by the Royal Society. In 1662, he published *The Art of Graveing and Etching, wherein is exprest the true way of Graveing in Copper*, which was "sold at his Shop next to ye Signe of ye Drake without Temple Barr."[16] It is this text, I argue, that helped to codify the standards of accuracy that developed over the course of the seventeenth century. Faithorne, in *The Art of Graveing and Etching*, provided his proven method for creating accurate copies of others' drawings that would outlast the fleeting observations of a single person.

Reading Faithorne's text alongside those of Robert Boyle (1627–1691) affords a better understanding of the development of standards of accuracy in the second half of the seventeenth century. Boyle is an exemplary figure because of his noted commitment to experimentation, which is borne out in his published works, as well as for the prominent role he played as a gentleman natural philosopher in mid-century London.[17] Boyle professed a preference for a plainer style of writing, to more clearly inform his readers.[18] In addition to being a founding member of the Royal Society, Boyle wrote many texts promoting experimentation and explained in great detail how to conduct experiments.[19] Boyle's writings, like Faithorne's, stress the necessity of careful and precise actions in order to ensure accuracy in the methods they espoused. In addition, I have chosen to study Robert Boyle's writings because of his connection to William Faithorne. In autumn 1664, William Faithorne was working on a portrait of Robert Boyle.[20] It was among the many Faithorne did of Fellows of the Royal Society and was one of the many projects that tied him to the Royal Society. In addition to a traditional three-quarter bust, the portrait includes a rendering of one of Boyle's air-pumps. The portrait of Boyle, Faithorne's writings, and Boyle's writings all serve as testaments to the importance of accuracy within the network of individuals working in and around the Royal Society, as all three visually and verbally embody the importance of careful work in the creation of knowledge.

Before Abraham Bosse (1602–1676) published his *Traicté des manières de graveur en taille douce* [Treatise on the manner of copper-plate engraving (Paris, 1645)], there were no texts dedicated solely to the description of engraving and etching.[21] Prior to his treatise, books that described the arts of painting and drawing merely included brief sections on engraving and etching.[22] Bosse was writing his manual in an attempt to shore up the place of intaglio printing in the hierarchy of arts and to gain acceptance for printmaking in French academic circles. Faithorne's 1662 translation of Bosse's text, by contrast, was not concerned with the academic standing of engraving in England. Although Faithorne and Bosse knew one another and had shared concerns about the regulations placed on printmakers and print-sellers in Paris, their reasons for publishing a manual on the arts of etching and engraving in their native languages were quite different.[23] The differences between their goals for publishing a manual come through most clearly in their paratexts. In his preface, Bosse sets up engraving and etching as ancient arts that have played important roles in human history since Moses came down from the Mount.[24] Faithorne, on the other hand, quickly dismisses the need to justify his choice of occupation: "I shall not trouble my self to speak in the Commendation of this Art." Rather than providing a justification of his art, Faithorne is concerned "to present my Countrey with something of use, profit, and delight." Instead of using his text to aggrandize the place of engraving in the hierarchy of the arts, as Bosse did, Faithorne uses his work for the benefit of his countrymen and to educate them about his art. Faithorne's book is grounded in an English context, and he gives instructions that are specific to English artists.[25]

Although Faithorne "used him [Bosse] as an Author in this Work," he is quick to point out in his letter "To the Lovers of this Art," at the beginning of the book, that "I have not traced him so closely as to make it a meer Translation; but added something, making use of what I thought necessary."[26] Stylistically, Faithorne departs from Bosse's text; that is, Faithorne's prose is plain and direct, whereas Bosse's is more florid and convoluted. Faithorne more clearly delineates the steps that need to be taken at various stages of the process. His major changes are organizational. Whereas Bosse's directions for how to transfer an image onto a plate are split into two in the section on etching, Faithorne combines these two sections into one and places it at the beginning of the section "On Graving."[27] This change makes engraving the primary locus for considerations of methods for transferring images onto the plate, shifting the location from Bosse's preferred method of etching to Faithorne's engraving. Faithorne's book is not a "meer Translation"; he has streamlined Bosse's prose, reorganized sections, and then added information pertinent to an English audience.

My claim is that Faithorne's method is structurally and rhetorically similar to the one outlined by members of the Royal Society for the successful practice of experimental natural philosophy. Faithorne, like the Fellows of the Royal Society, was working toward developing a clear and coherent technical language that stood in marked contrast with the high style of Bosse and those enmeshed in the Aristotelian tradition.[28] The following sections assert that Faithorne is working with a language of accuracy. He and Fellows of the Royal Society, with whom he was connected through networks of exchange and collaboration, were actively creating a new way of writing about accuracy through the production of standards. I am calling the rhetorical tools used to articulate these standards a language of accuracy, and this development is homologous to development of the concept of accuracy itself.

Language has been defined as "the body of words and the systems applying to their use that are common to a people of the same community, area, nation, or culture."[29] In this case, the community that shared this language is the social network of the Fellows and the artisans they were working with to publish their research and findings. The systems of this language of accuracy I am concerned with are its grammar and syntax. The lexicon, or body of words, used in this language is also important for the creation of standards of accuracy in the circle of the Royal Society because the words repeated in these texts emphatically highlight the attempts made to control, manage, and produce accuracy.

In its broadest sense, grammar is defined as "a statement, or series of statements, of the way a language works." This language of accuracy works through a series of best practices. Faithorne developed these best practices using other texts on engraving as well as his own experience as an engraver. He provided the reader with large-scale rules and structures that guide the creation of an accurate product. The grammar of accuracy that he deployed thus furnished a clear and coherent method to be followed. In addition to a grammar of best practices, this language of accuracy also adheres to a syntax that governs how each step in the process is described. In terms of linguistics, syntax is defined as "a study of the rules for forming grammatical sentences in a language, the arrangements of morphemes, words, and phrases in meaningful utterances," or as "the structure of sentences." Within this language of accuracy, the syntax works to nuance the method laid out by the grammatical best practices. That is, the syntax governs how the grammar/method is put into practice. In literary terms, "lexicon" is defined as "the vocabulary of a particular language, activity, social class, etc."[30] The social network that existed between artisans and Fellows shared this lexicon of accuracy. Within this language, there is a set of words that are continually repeated, particularly at points of importance for understanding the best practices.

It is worth noting here that accuracy is not among the most frequently used words in the lexicon of this language I am describing. The etymology of "accurate" given in the *Oxford English Dictionary* traces the word back to the Latin word "accuratus," meaning "performed with care, exact."[31] The words "care" and "exact" are among the most prominent in this lexicon. "Accurate" occurs once in Faithorne's text and is closely aligned with the importance of "neatness," and both are understood as the results of great care, tying his usage to those of Fellows of the Royal Society:

> Be *carefull* that your fingers do not interpose between the plate and the Graver, for they will be troublesome, and hinder you in carrying your Graver level with the plate, so that you cannot make your strokes with that freedome and *neatnesse*, as otherwise you may. This I think fit to give you notice of in this place, because the skill of holding your Graver is that which you must first *perfectly* learn, and be able to practise without pain or difficulty; or else you will not gain so *great* a *readinesse and command of your hand*, as is required in *an accurate and skilfull Graver* (emphasis added).[32]

I have quoted this passage at length to give a sense of the range of meanings associated with accuracy in this text and the importance of methodical work in the achievement of success. In this passage, the term "graver" is used to refer to both the tool used by an engraver and the engraver himself. This ambiguity reinforces the sense that the reliability and truth of testimony were related to an individual and, in this case, one who was trained to be accurate and skillful. This collapse of the artisan and his tool allows us to understand *The Art of Graveing and Etching* as a manual for making *and* being.

Writings in the language of accuracy, such as Robert Boyle's, helped to solidify the best method for conducting experiments, just as the language of accuracy worked in Faithorne's text to control the production of printed illustrations. Both experiments and "accurate" renderings must stay close to the object of study, convey or even produce visual data or information reliably, and be reproducible. Engravings produced repeatable results, in the sense of multiples that can circulate and through conventions of representation that can be replicated by others. Looking closely at how William Faithorne and Robert Boyle used the language of accuracy in their writings allows for a clearer understanding of the development of standards of accuracy in seventeenth-century London.

The Grammar of Accuracy: An Engraver's Best Practices

As the first text in English to outline a detailed method for engraving and etching, Faithorne's *The Art of Graveing and Etching* provides the reader with a set of best practices to be followed carefully and closely. These established the possibility that an engraver could reproduce a design as it was given to him and served as the large-scale structure for explicating a method for producing an accurate engraving or etching. By looking closely at the section on engraving, in particular, we begin to understand the structure of the method proposed in Faithorne's volume.

This portion of the book is divided into four sections: "the several wayes of drawing the design upon the plate," "the forms of graving Tooles, as also the manner of whetting your Graver," "the manner how to hold your Graver," and finally "the manner of governing your hand in Graving."[33] These four sections detail the steps to be taken to create an engraving; beginning with transferring a design onto the plate and ending with carving the desired lines into it. In each of these sections, the best method for accomplishing every step is described and the reader is provided with tips about possible problems that might occur and how to avoid or correct them. Looking at the section on how to hold the burin will help us understand the stress Faithorne puts on heeding his advice and proceeding in the order he promotes as well as lend insight into the method for producing accuracy.

Once the design has been transferred to the plate and the burin is exactly whetted, Faithorne indicates that it is time to learn the best practices for holding the burin. Faithorne judges that his reader is now ready to learn the next crucial step in attempting to make an accurate engraving. Here he stresses the necessity of proceeding in the mandated order.

"This I think fit to give you notice of *in this place*, because the skill of holding your Graver is that which you must first perfectly learn, and be able to practise without pain or difficulty; or else you will not gain so great a readiness and command of your hand, as is required in an accurate and skilfull Graver" (emphasis added).[34] Before a single line is cut into the copper, the rules governing the handling of the burin must be learned. As noted, in this passage the distance between the instrument and the man collapses. The grammar that structured how the graver is to be used now extends to the artisan. Not only are actions to be perfected, but the artisan is as well. If you are to be "an accurate and skilfull Graver," you must accept and follow the logic and order of the best practices. Becoming a "skilfull Graver" allows for the repeated production of accurate representation and helps to reduce variations in quality over time.

The care Faithorne exhibits in outlining the best practices or ideal method for creating an engraving is mirrored in the level of detail Robert Boyle provides in his description of how to construct an air-pump like the one shown in Faithorne's portrait of him.[35] The inclusion of this device in the background of Faithorne's portrait of Boyle indicates that it was seen as emblematic of Boyle's success and renown in 1664 as well. Growing out of the tradition established by Anthony Van Dyck (1599–1641) and his contemporaries, portraits of great men often included signs of their status, be it their land holdings, their expansive homes, or their travels.[36] In Robert Boyle's case, this tradition is adapted to accommodate a new signifier of status. The middle distance of the portrait does not contain a long vista of open countryside or a ship coming into port; instead it features a machine designed to unlock the secrets of nature.[37]

ROBERTVS BOYLE ARM:

Figure 1. William Faithorne, *Portrait of Robert Boyle*, 1664. (Reproduction by permission of the Syndics of The Fitzwilliam Museum, Cambridge)

In an attempt to ensure that his readers would be able to recreate the air-pump he described, Robert Boyle provided detailed instructions at the beginning of his *New Experiments Physico-Mechanicall, Touching the Spring of the Air and Its Effects* for how to construct one. For this argument, I am less interested in the fact that people had trouble recreating the air-pump than I am in the level of detail provided in the method of constructing one.[38] After outlining some of the background that led to his trials with making a vacuum chamber, Boyle describes the best practices for constructing an air-pump that will keep the air out of the chamber for short periods of time. The method for assembling an air-pump is structured by brief descriptions of a section of the machine followed by detailed descriptions of how to construct each piece.

In the case of the pump, a brief description is provided ("The Pump consists of four parts, a hollow Cylindre, a Sucker, a handle to move that Sucker, and a Valve.") and then the method for building each piece is provided. The level of detail provided in Boyle's best practices is so high that the next two and a half pages of text are given over to describing these four parts. Here is an excerpt from the account of how the cylinder is made: "The Cylindre was (by a pattern) cast of brass; it is in length about 14 inches, thick enough to be very strong, notwithstanding the Cylindrical cavity left within it; this cavity is about three inches in Diameter, and as exact a Cylindre as the Artificer was able to bore."[39] A succinct definition ("a hollow Cylindre") is expanded upon to ensure that the reader can create a replica of what Boyle is describing. As this brief example shows, both artisans and members of the Royal Society were concerned with outlining the best practices for creating standardized outcomes, whether two- or three-dimensional. In both cases, the best practices are dependent on the demonstration of care in the production of the objects of study.

The Syntax of Accuracy: Regulating the Graver

The best practices Faithorne outlined throughout *The Art of Graveing and Etching* function as a grammar of accuracy in so far as they provide a structure around which a method for creating an accurate rendering is built. The ways Faithorne described and clarified the best practices function as the syntax. This section discusses elements of this syntax and how they work to assure the level of accuracy demanded by natural philosophers and natural historians. The key components of this syntax of accuracy include: the use of conditional statements to clarify how to know if a best practice has been properly implemented; the inclusion of cautionary asides that outline how the engraver might go astray in the process of making an engraving; and the framing of descriptions of best practices with references to the plates included in the book. These three components work to regulate the actions of the engraver and standardize his output.

The principal syntactical element I want to discuss in greater detail is the framing of best practices with references to the images interleaved with the text. There are ten plates included in *The Art of Graveing and Etching*, all of which William Faithorne engraved. These figures are referred to throughout the text. "Figure(s)" occurs thirty-three times in the book, indicating that the individual images are referred to multiple times. References to the figures in the text perform a similar function as the other two syntactical elements, as they work to clarify the reader's understanding of the method and further ensure his ability to replicate the method in his own work. The images are based on Faithorne's own experience, as he had to put his own method into action to create these prints that accurately translate his verbal method into pictorial evidence. If this sounds like circular logic, that is part of the point: the plates and the text are both the products of Faithorne's years of practical experience as an engraver and work to reinforce his credibility as someone capable of providing a method worth following. The relationship between image and text reinscribes the collapse of the engraver and his tools and further stresses the need for careful actions.

The text facing the last plate begins with a reference to the image and continually refers the reader back to the image. Section 27, "The manner how to hold your Graver, with other particulars," begins by directing the reader to the facing image: "You may see also that the uppermost part of this figure describes to you the form of the two Gravers, with their handles fitted for the whetting."[40] Two sets of information can be gleaned from the upper portion of this image at this moment. First, the reader is shown another view of what the square and "lozeng" points of a burin look like. These two possibilities for the point were shown in the previous image from a different angle. Second, this section of the image shows how engravers will often cut off half of the handle to make sure that the knob does not rub against the plate. The figure both provides more detailed information than the text and models behavior; in this case, we can assume Faithorne is using a burin that has been cut in the manner he advocates.

Although the reader is directed to the image to learn about these two aspects of preparing the burin for engraving, the reason for cutting the handle is only given after the reference to the image: "They that use this Art, do before they make use of them, commonly cut away the part of the knob or bowl that is at the end of their handles, which is upon the same line with the edge of their Graver; to the end it may not obstruct or hinder them in their graving, as the figure II. shews you."[41] This additional information, which clarifies what is seen in the upper portion of the image, is coupled with a reference to another part of the same image. The first paragraph of this section (which I have quoted in its entirety in the previous two quotations) begins and ends by directing the reader to the facing image. The best practices for preparing the burin rely on visual information to convey clearly the how and why of the method.

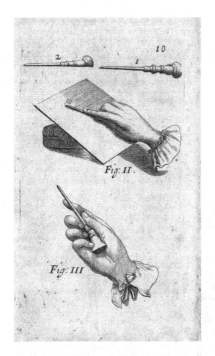

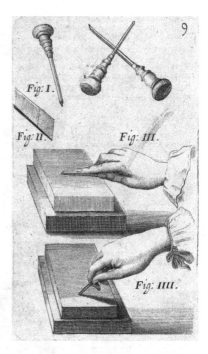

Figure 2. William Faithorne, plate 10, *The Art of Graveing and Etching*, 1662. (Reproduced by permission of The Huntington Library, San Marino, California)

Figure 3. William Faithorne, plate 9, *The Art of Graveing and Etching*, 1662. (Reproduced by permission of The Huntington Library, San Marino, California)

Now that the "other particulars" associated with holding the burin have been textually and visually explained, the third paragraph of this section and the third portion of the image deal more directly with holding the burin:

> The third figure describes to you the way of holding the Graver; which is in this manner. You must place the knob or ball of the handle of your Graver in the hollow of your hand, and having extended your forefinger towards the point of your Graver, laying it opposite to the edge that should cut the copper, place your other fingers on the side of your handle, and your thumb on the other side of the Graver, in such sort that you may guide your Graver flat and parallel with the plate; as you may see in the IIII. [III] figure.[42]

The reader is to first look at the image, then read the description, then look back at the image to be sure that he understands how best to hold the burin. This visual reinforcement of the best practices Faithorne outlines works to ensure that his readers develop a careful and accurate practice. The plates in *The Art of Graveing and Etching* visually embody Faithorne's exposition of the ideal method of producing an engraving and model behavior for his readers.

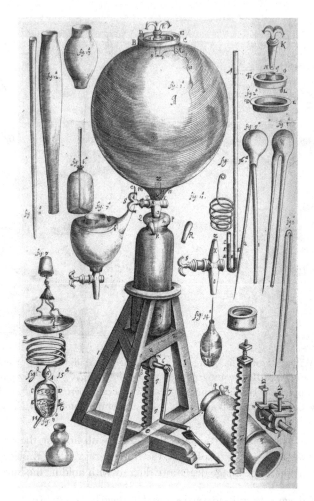

Figure 4. From Robert Boyle, facing page 1, *New Experiments Physico-Mechanicall, Touching the Spring of the Air*, 1662. (Reproduced by permission of The Huntington Library, San Marino, California)

In a text on experimental natural philosophy, the images included model the ideal construction of the experimental set-up. As with *The Art of Graveing and Etching*, Boyle's *New Experiments Physico-Mechanicall* frames descriptions of best practices with reference to images to help clarify the steps to be taken. Although there is only one plate accompanying the text, it is made up of a series of figures, each with its own set of labels. The first reference to this figure occurs on page nine, where Boyle tried to clarify the shape of the glass that will serve as the vacuum chamber. The text gives little sense of the ideal shape: "[The glass vessel] consists of a Glass with a wide hole at the top, of a cover to that hole, and of a stop-cock fastened to the end of the neck, as the bottom."[43] To know that the vessel should be roughly oval in shape and taper at the end that will connect to the stop-cock, it is necessary to refer to the provided figure. This carefully engraved image is covered with letters and numbers that key into the accompanying text, and each line is precisely cut into the copper to ensure clear communication. In both Boyle's writings and Faithorne's *The Art of Graveing*, the references to the accompanying plates work with the other syntactical elements outlined to further refine the reader's attempts to put the method described into practice.

The Lexicon of Accuracy: Exactly the Right Words

Where the grammar and syntax of accuracy work on a macro level to develop standards of accuracy, the lexicon of this language works on a micro scale. Embedded in each of the passages I have discussed are individual words that add emphasis and further highlight important components. This section discusses two crucial elements of the lexicon of accuracy: the vocabulary of actions and that of outcomes. In both cases, the emphasized words serve as markers of an underlying interest in accuracy. Some of the words that stress the importance of accuracy in actions are "care/ful," "great," "very," and "must." The engraver's actions, like those of the scientist, *must* be careful and well thought out and his equipment well chosen. The outcomes of accurate actions are expected to be neat and exact. Throughout *The Art of Graveing and Etching* and the writing of Robert Boyle, these clusters of words work on a micro level to create standards of accuracy.

The vocabulary of actions appears throughout Faithorne's text. The repetition of these words controls the production of the engraver's tools as well as his actions more generally. It is through their repetition at crucial junctures that these words help to solidify the standards of accuracy. In the forty-eight pages of text that comprise *The Art of Graveing*, the word "care" appears fourteen times and "carefull" thirteen times.[44] In many of these instances, "care/ful" is modified by "great," "very," "[e]special," or some combination therein. Of the thirteen times that the word "great" appears, it modifies "care" five of those times. Every time "carefull" appears in the text, the reader is told to "be carefull" in performing a certain task.[45] In some of the instances being careful is further emphasized, as in: "you must be very carefull," "be carefull above all," and "be carefull in especiall manner." The would-be engraver is being fashioned as a more careful craftsman and also has to be a careful reader in order to notice and recognize the instances in which care is of the utmost importance. In addition to being careful, the engraver must also "take great care" with his work and "have an especiall care" in his work.[46] Care then is both something to embody and to use further exemplifying the way this is a manual for being and making. In the seventeenth century, accuracy was coming into the English language and was closely aligned with "care" as it was derived from the Latin word *cura*, meaning "care taken."[47] The high frequency of "care/ful," therefore, indicates a heightened attention to accurate actions.

In addition, the verbs Faithorne uses are emphatic. "Must" appears fifty-two times, whereas "should" only occurs nineteen times and "might" only five times. The reader is not told that he might try this method or that but rather that he must perform a certain action the way Faithorne states. Faithorne was not outlining a method by which users could pick and choose the parts that fit with their style, rather *The Art of Graveing* works to create standards of accuracy that can only be achieved by closely following the best practices provided and not deviating from the prescribed method. An engraver must have "a very great care" in his actions to produce tools and prints that are exact and neat in their execution.

"Exact," "exactly," and "exactnesse" occur two, five, and one times respectively, emphasizing the need for precision.[48] Exactitude is particularly prized in the whetting of the needle and burin. Twice in the description of "how to guide your needles upon the plate," the necessity of having properly "whet [it] exactly round" is emphasized. The phrase, "whet exactly round," occurs in the context of both a conditional statement and a cautionary aside.[49] Here, the lexicon of accuracy stresses the need to follow the method for whetting the etching needle previously outlined in order to cut the varnish accurately. A syntactical clarification rests on the importance of lexical stress to emphasize the importance of following the best practices.

While "exact(ly)" is used to qualify desired outcomes, neatness, in and of itself, is a desired outcome. When "neatness" occurs in the text, it is usually paired with another term (such as "clearnesse," "firmnesse," "curiosity," or "command") that clarifies and augments its meaning. Over the course of the seven instances of "neatnesse" in the text, the term takes on a range of meanings. The engraver's work should possess "neatnesse and clearnesse"; indicating that neatness is akin to clarity, an essential trait for the accurate rendering of the natural world.[50] Twice "neatnesse" is paired with "curiosity."[51] In these passages, the twinned dexterity of the hand and the intellect are desired and necessary in the art of engraving. Neatness is then to be expressed both in the lines carved into the plate and in the overall working habits of the engraver. He is to both be neat and make images that are neat and therefore clearly and cleverly convey their meaning. The lexicon of accuracy adds nuance on a micro scale to both the actions the engraver is expected to perform and the resulting products. The same level of care was necessary in the set-up of experiments members of the Royal Society performed, and a similar lexicon was used.

Robert Boyle was equally prescriptive in his use of language in his writings on the best practices for constructing an air-pump. As with Faithorne's lexicon of accuracy, Boyle urges his readers to work with great care, and exactness is again expected. Boyle also stresses that he has proceeded carefully in his own experiments and that his text relates the care he took with building his air-pump: "That foreseeing that such a trouble as I met with in making those tryals carefully, and the great expence of time that they necessarily require (not to mention the charges of making the Engine, and imploying a man to manage it) will probably keep most men from trying again those Experiments; I thought I might doe the generality of my Readers no unacceptable peice of service, by so punctually relating what I carefully observ'd."[52] He both worked and observed carefully, and the very care he took in doing these things made his text of use to others. Over the course of more than 400 pages of text, "care," "careful," and "carefully" occur eight, two, and thirty-two times, again emphasizing the importance of acting with care. As was the case in *The Art of Graveing*, "very" and "great" are used over and over again to add emphasis, with "very" appearing 250 times and "great" 133.

Exactitude is also a prized characteristic of the outcomes of careful actions in Boyle's description of his ideal method for constructing an air-pump. "Exact," in various forms, occurs thirty-six times in the text.[53] The stop-cock of the pump is designed for "exacter exclusion of the Air." The hole in the top of the cylinder is blocked with a piece of leather "that it may the more exactly hinder the Air from insinuating it self."[54] These two examples show that the lexicon of accuracy is constantly working to highlight and nuance the best practices that Boyle presents and that exactitude is essential to creating a vacuum chamber. By being careful and exactly replicating best practices, a would-be engraver or scientist could attempt to reproduce a drawing or an experiment.

Adoption and Acceptance: Some Conclusions

The language of accuracy used by Faithorne was both adopted by other artisans writing on the practice of engraving and accepted by members of the Royal Society. These two groups were both parts of the "Lovers of this Art" to whom Faithorne appealed at the beginning of his book. In his opening epistle, Faithorne writes that he has decided to publish this book on the art of engraving and etching now "because it hath arrived to such an height in these our latter times, as it becomes a fit subject for our Kingdomes knowledge and practice."[55] Both knowledge and practice are here stressed. *The Art of Graveing and Etching* is not just a technical manual aimed at practitioners but also a treatise meant to increase knowledge of these arts to a broader audience. Faithorne was able to reach both of these groups with his text, as is evidenced by the adoption of his text as a standard to be emulated by fellow practitioners and as a definitive text

on the subject by Fellows of the Royal Society who were interested in gaining
theoretical knowledge of technical subjects.

Examples from books produced after the publication of *The Art of Graveing
and Etching* shed telling light on the impact of Faithorne's text. *The Excellency of
the Pen and Pencil* (London, 1668) and William Salmon's *Polygraphice* (London,
1672) provide proof that Faithorne's best practices became the industry standard
and that his language of accuracy prevailed in technical manuals. While these
texts do not list Faithorne as an author, it is clear his text was consulted and
relied on heavily in the writing of these books. Sections of Faithorne's description
of the best practices to follow for sharpening the burin are lifted wholesale.

> From *The Excellency of the Pen and Pencil*: It is very necessary that you take great
> care in the exact whetting of your Graver; for it is impossible that you should
> work with that neatness and curiosity as you desire, if your Graver be not very
> good, and rightly whetted.[56]
> From *Polygraphice*: It is impossible that the work should be with the neatness
> and curiosity desired, if the grave not be onely very good, but also exactly and
> carefully whetted.[57]
> And from *The Art of Graveing and Etching*: It is very necessary that you take
> great care in the exact whetting of your Graver; for it is impossible that you
> should ever work with that neatnesse and curiosity as you desire, if your Graver
> be not very good, and rightly whetted.[58]

The Excellency of the Pen and Pencil does not list an author on its title page, instead
it says that the following text has been "Collected from the Writings of the
ablest Masters both Antient and Modern."[59] It is clear from this example that
Faithorne's contemporaries considered him an able master. Although Salmon
had changed the word order slightly, the sentiment is the same, and the frequency
of words associated with the lexicon of accuracy has increased.

While these two examples show the immediate impact Faithorne's text had
on published methods for the making of intaglio prints, his legacy among artists
extended well into the eighteenth century as well. Not only was his text reissued
in 1702, but it also became the model for later engraving manuals. In 1747,
*Sculptura historico-technica: Or, The history and art of Ingraving . . . Extracted from
Baldinucci, Florent le Compte, Fairthorne, the Abecedario pittorico, and other authors*
appeared on the London market. This book included technical instructions for
engraving and etching based primarily on Faithorne's *The Art of Graveing and
Etching*. In addition, the plates in *Sculptura historico-technica* are based entirely on
Faithorne's plates. *Sculptura historico-technica* went through two further editions,
in 1766 and 1770, attesting to the lasting market for Faithorne's treatise on the
art of engraving.

The continued presence of Faithorne's text on the London market demonstrates artists' interest in his book and their adoption of his method. Members
of the Royal Society also valued Faithorne's efforts. In 1662, John Evelyn, a

founding member of the society, published *Sculptura: or The History, and Art of Chalcography and Engraving in Copper*. In "An Advertisement" at the end of *Sculptura*, Evelyn said of Bosse's book and Faithorne's translation of it: "It was my intention to have added it [a translation of Bosse's text] to this *History* of mine, as what would have render'd it a more accomplish'd Piece; but, understanding it to be also the design of Mr. *Faithorn*, who had (it seems) translated the first part of it, and is himself by Profession a *Graver*, and an excellent *Artist*; that I might neither anticipate the *Worlds* exspectation, nor the *Workmans* pains, to their prejudice, I desisted from printing my *Copy*, and subjoyning it to this *discourse*."[60] According to Evelyn, Faithorne's status as a professional engraver and "an excellent *Artist*" made him more qualified to translate Bosse's text and present it to an English-speaking audience. While Evelyn said "it seems" that Faithorne had produced a translation, he did own a copy of the book and happened to visit Faithorne's shop days before *The Art of Graveing* was published.[61] Evelyn went on to state that it was his hope that others would follow Faithorne's lead and uncover the mysteries of their arts to a wider audience for the benefit of PHILOSOPHY (in all caps). Evelyn's comments about Faithorne's book and his ownership of a copy indicate that the standards Faithorne provided were accepted by a prominent member of the Royal Society who was particularly concerned with the history of trades. The adoption of Faithorne's language by artisans and its acceptance as appropriate to the goals of the Royal Society indicate that *The Art of Graving and Etching* had a lasting impact on both segments of its intended audience.

This analysis of the reception of Faithorne's text shows that the production standards he introduced to English printmaking were considered important both to those who would be including engraved illustrations in their books and those who were publishing manuals. For both groups, Faithorne was establishing a clear and coherent method that helped them to believe it was possible to accurately translate a drawing into a print. Following this exacting method, an engraver could create images that would pass for transparent representations of nature. In so doing, they worked to erase the signs of their intervention in the process of creating visual knowledge that circulated in printed books. Faithorne's text became worth emulating in part, I want to suggest, because of his multi-layered connections with the Fellows of the Royal Society. His ability to create likenesses of them, skill in accurately rendering their machines, and reputation as a keen observer all contributed to the longevity of his method and the prevalence of his language of accuracy in writings on the art of engraving.

Notes

1. My research indicates that engraving was used as a generic term in the period, and engraving and etching were seen as complementary processes. For example, William Faithorne places "graveing" before etching in the title of his engraving manual,

The Art of Graveing and Etching, even though three quarters of the text is dedicated to the chemical intricacies that must be mastered with etching. In addition, the section on etching is titled "The Art of Graving with *Aqua fortis*" (William Faithorne, *The Art of Graveing and Etching wherein is exprest the true way of Graveing in Copper* [London, 1662], 30). Throughout this essay, in keeping with period usage, I will use "engraving" as the generic term for the intaglio processes prevalent in the seventeenth century.

2. While my work is indebted to that of Lorraine Daston and Peter Galison, I am not arguing that images produced for the Royal Society during the seventeenth century were "objective" in the same way that those produced for scientific atlases during the eighteenth and nineteenth centuries were. I do not see the same emphasis on "scientific objectivity" in the seventeenth century that Daston and Galison found in their research on a later period. Lorraine Daston and Peter Galison, "The Image of Objectivity," *Representations* 40 (1992): 81–128; Lorraine Daston and Peter Galison, *Objectivity* (New York: Zone, 2007).

3. This line of my argument draws on the work of Steven Shapin; see his *The Social History of Truth: Civility and Science in Seventeenth-Century England* (Chicago: University of Chicago Press, 1994).

4. The importance of civility and gentlemanly conduct to the inner workings of the Royal Society has been much discussed in the literature on the Royal Society. See for instance: Shapin, *Social History of Truth*; Steven Shapin and Simon Schaffer, *Leviathan and the Air-Pump: Hobbes, Boyle, and the Experimental Life* (Princeton, N.J.: Princeton University Press, 1985); and Peter Dear, "*Totius in Verba*: Rhetoric and Authority in the Early Royal Society," *Isis* 76, no. 2 (1985): 144–61.

5. Barbara Shapiro writes that the "fact" did not begin as something that existed in nature but instead originated in the field of law. Barbara J. Shapiro, *A Culture of Fact: England, 1550–1720* (Ithaca, N.Y.: Cornell University Press, 2003), 105. In addition, Mary Poovey draws attention to the role "interest" played in the creation of the modern fact in early modern England and writes that members of the Royal Society were concerned with the uninterested accumulation of knowledge and worked to separate judgment from knowledge collection, which was in contrast to Thomas Hobbes's more neutral definition of "interest." Mary Poovey, *A History of the Modern Fact: Problems of Knowledge in the Sciences of Wealth and Society* (Chicago: University of Chicago Press, 1998), 93.

6. This essay is not an argument for a sudden revolutionary change in the way science was practiced. Rather I am arguing that studying the development of standards of production for printed images, which helped build consent surrounding what was an accurate image, further refines our understanding of how scientific inquiry changed during the seventeenth century. I do not intend in this essay to enter into the fraught territory of whether or not there was a Scientific Revolution, nor do I intend to rehearse the historiography of the topic. Instead, I want to press on the assertion that improvements in printmaking affected the nature of research in the seventeenth century. In their introduction to the Cambridge History of Science volume *Early Modern Science,* Lorraine Daston and Katharine Park chose to forego the phrase entirely and instead use "early modern science" to demarcate the period under consideration. They go so far as to say that the "Scientific Revolution" is a mythology. Katharine Park and Lorraine Daston, eds., *Early Modern Science* (Cambridge: Cambridge University Press, 2006), 15.

7. The phrase "experimental learning" is taken from the title of Marie Boas Hall's book. She in turn takes it from the minutes of the Royal Society. Marie Boas

Hall, *Promoting Experimental Learning: Experiment and the Royal Society, 1660–1727* (Cambridge: Cambridge University Press, 1991).

8. Margery Purver, *The Royal Society: Concept and Creation* (Cambridge, Mass.: MIT Press, 1967), esp. chap. 3, "A New System of Natural Philosophy II, The Royal Society." The literature on the place of experiment in the workings of the Royal Society is extensive. In addition to the writings of Marie Boas Hall and Shapin and Schaffer, Michael Hunter's work on this topic is also useful: Michael Hunter, *Establishing the New Science: The Experience of the Early Royal Society* (Woodbridge, Suffolk: Boydell, 1989).

9. Thomas Sprat, *The History of the Royal Society of London for the Improving of Natural Knowledge* (London, 1667), 83.

10. The first use of the word recorded in the *Oxford English Dictionary* is in Henry More's *An antidote against atheisme*. The *OED* gives the date for this occurrence as 1662, the same year that More was elected as a Fellow of the Royal Society. Members of the Royal Society are also associated with the early uses of the words "accurate" and "accurateness." *OED*, s.v. "accuracy," "accurate," and "accurateness."

11. I am by no means the first to be interested in the transmission of knowledge through images. See for instance, Sachiko Kusukawa and Ian Maclean, eds., *Transmitting Knowledge: Words, Images, and Instruments in Early Modern Europe* (Oxford: Oxford University Press, 2006).

12. The Fellows who sat for Faithorne are John Aubrey, Robert Boyle, Edmund Castell, Francis Glisson, Edmond King, Charles Leigh, Henry More, John Ray, and John Wallis. This list is based on Bell and Poole's list of portraits that were done from life. C. F. Bell and Rachel Poole, "English Seventeenth-Century Portrait Drawings in Oxford Collections: Part II, Miniature Portraiture in Black Lead and Indian Ink," *Walpole Society* 14 (1925–26): 52–54. Their list has been cross-referenced with the Royal Society's online database of Fellows.

13. Of the six signed plates, Faithorne signed two.

14. This plate is a large map that accompanies the article "A Narrative of the Conjunction of the two Seas, the Ocean and the Mediterranean, by a Channel, cut out through Languedoc in France, by the Authority of the Most Christian King Louys XIV, and the Contrivance and Management of Monsieur Riquet: Englished by the Publisher out of the French, lately sent to the same from Paris, together with the printed Mapp of the said Channel, here contracted and explained," *Philosophical Transactions of the Royal Society* 4, no. 56 (February 17, 1670): 1123–28.

15. John Brown, "A Remarkable Account of a *Liver*, appearing *Glandulous* to the Eye; communicated by *Mr. John Brown*, Chirurgeon of *St. Thomas's* Hostpitall in *Southwark*; in a Letter to one of the *Secretarys* of the *Royal Society*," *The Philosophical Transactions of the Royal Society* 15, no. 178 (1685):1266–68. The plate was engraved by "M. Burgesses."

16. Faithorne, *Art of Graveing and Etching*, title page.

17. The importance of Robert Boyle to the history of the Royal Society and the development of modern science has been made clear in the work of Steven Shapin, Simon Schaffer, and especially Michael Hunter. See for example: Shapin, *Social History of Truth*, where Boyle is set up as the quintessential gentleman-philosopher; Shapin and Schaffer, *Leviathan and the Air Pump*; and the writings of Michael Hunter, in particular, *Robert Boyle (1627–1691): Scrupulosity and Science* (Woodbridge, Suffolk: Boydell, 2000) and *Robert Boyle Reconsidered*, ed. Michael Hunter (Cambridge: Cambridge University Press, 1994).

18. "And first, as for the style of our Experimental Essaies, I suppose you will readily find that I have endeavour'd to write rather in a Philosophical than a Rhetorical strain, as desiring that my Expressions should be rather clear and significant than curiously adorn'd. . . . And certainly these Discourses, where our Designe is only to inform Readers, not to delight or perswade them, Perspicuity ought to be esteem'd at least one of the best Qualifications of a style." Robert Boyle, *Certain Philosophical Essays* (London, 1661), 11. John Harwood writes about the balance Boyle tried to find in his writing between plain style and the tools of classical rhetoric he thought were useful for engaging readers. John T. Harwood, "Science Writing and Writing Science: Boyle and Rhetorical Theory," in *Boyle Reconsidered*, 46.

19. See for example: Robert Boyle, *New Experiments Physico-Mechanicall, Touching the Spring of the Air* (Oxford, 1660); Boyle, *Certain Physicological Essays* (London, 1661); Boyle, *Some Considerations Touching the Usefulnesse of Experimental Natural Philosophy* (Oxford, 1664); and Boyle, *A Continuation of New Experiments Physico-Mechanicall, Touching the Spring and Weight of the Air and Its Effects* (Oxford, 1669).

20. Based on the correspondence between Robert Boyle and Robert Hooke, acting as Boyle's agent, we can date this portrait to the fall of 1664. *Correspondence of Robert Boyle*, ed. Michael Hunter, Antonio Clericuzio, and Lawrence M. Principe, vol. 2, *1662–5* (London: Pickering & Chatto, 2001), 304, 316, 412, 442.

21. Linda C. Hults, *The Print in the Western World: An Introductory History* (Madison: University of Wisconsin Press, 1996), 212, 285.

22. An example of this trend is John Bate's *The Mysteryes of Nature and Art Conteined in four severall Tretises* (London, 1634). In this text, the art of engraving is lumped in with a wide range of other mysteries Bate is going to reveal to his readers.

23. Marianne Grivel found a document that includes both Bosse and Faithorne's signatures on a petition protesting François Mansart's 1651 proposed tax on prints. Marianne Grivel, *Le Commerce de l'Estampe à Paris au XVIIe Siècle* (Geneva: Librairie Droz, 1986), 94–96.

24. Bosse writes "ni qu'il est des plus anciens puis que Moyse en a escrit ainsi que d'une chose laquelle estroit de son temps fort en usage." Abraham Bosse, *Traicté des manières de graveur en taille douce* (Paris, 1645), 1.

25. Regarding the "Englishness" of Faithorne's book, for instance, at the beginning of the section on "How to planish and polish your Plate" he states: "Here in *England* you must buy your Cooper ready forged from the Brasiers." Faithorne, *Art of Graveing and Etching*, A2r, Av, 4.

26. Ibid., A2r.

27. In Bosse's table of contents, the sections are listed as "La maniere de s'apprester pour desseigner, contretirer ou calquer son dessein sur la planche vernie" [The manner of preparing to draw, transfer, or trace the design onto the varnished plate] and "La maniere de contretirer ou calquer le dessein sur le verny" [The manner of transferring or tracing the design onto the varnish]. Bosse, *Traicté des manières de graveur*, 6. The Faithorne section is listed as "The several wayes of drawing the design upon the plate." Faithorne, *Art of Graveing and Etching*, A4v.

28. Margery Purver has written persuasively that the Royal Society was building on the work of Francis Bacon in its call for a technical language. Purver, *Royal Society*, 54, 237.

29. Harry Shaw, *Dictionary of Literary Terms* (New York: McGraw-Hill, 1972), 217.

30. Ibid., 176, 370, 220.

31. *OED*, s.v. "accurate," etymology.

32. Faithorne, *Art of Graveing and Etching*, 45–46.

33. Ibid., A4v. I am using the section of the book on engraving as an example of the grammar of accuracy in the book. This should not be read as an indication of a lack of such grammar in the sections on etching.

34. Ibid., 46.

35. In large part because of the work of Steven Shapin and Simon Schaffer, the air-pump has become emblematic of the advances in experimental research in the seventeenth century. Shapin and Schaffer, *Leviathan and the Air-Pump*.

36. Zirka Zaraemba Filipczak, "Reflections on Motifs in Van Dyck's Portraits," in *Anthony Van Dyck*, ed. Arthur K. Wheelock Jr. et al. (Washington, D.C.: National Gallery of Art, 1990), 59.

37. Shapin and Schaffer identify the air-pump in Faithorne's portrait as a representation of the first version of Boyle's pump. Shapin and Schaffer, *Levithan and the Air-Pump*, 257.

38. Shapin and Schaffer trace the "career" of the air-pump and find that it was fraught with difficulties, as there was rarely one that worked properly for very long. Ibid., 229–31.

39. Boyle, *New Experiments Physico-Mechanicall*, 12–13.

40. Ibid., 45.

41. Ibid.

42. Ibid. As there are only three sections to this image, we can assume that "IIII." is a typographical error.

43. Boyle, *New Experiments Physico-Mechanicall*, 9.

44. I used a shareware program called WordCounter to calculate the frequency of words in the text.

45. Some form of "be carefull" occurs on pages 7, 15, 16, 25, 29, 31, 33, 34, 38, 40, 42, 45, and 47.

46. Faithorne, *Art of Graveing and Etching*, 15, 29, 34, 7, 21, 29, 44, 11. Other formulations involving taking care include: "taking especiall care" (21), "the more care taken" (23), and "take speciall care" (39). "A very great care" (30), "much more care" (31), and "a speciall care" (31) are also to be had.

47. *OED*, s.v. "accurate," etymology, and *Cassell's Latin Dictionary*, s.v. "cura."

48. "Exact": Faithorne, *Art of Graveing and Etching*, 32, 44. "Exactly": 12 (twice), 38, 41, 44. "Exactness": 44.

49. Ibid., 12. "If you find that your point cuts not freely and smoothly, 'tis because it is not whet exactly round." "You may perceive from what is said, that those points which you intend to make use of, in graving with *Aqua fortis*, ought to be whet exactly round, that they may turn more freely upon the plate."

50. The phrase "neatnesse and clearnesse" occurs in the following passage: "I shall only think it not amisse to advise you by the way, that in making your strokes with your Oval points, you must hold them as upright and streight in your hand as you can, and accustome your self to strike your strokes firm and bold, for that will contribute very much to their neatnesse and clearnesse" (ibid., 15).

51. The first instance pertains to the process of coating the plate with a whitening agent instead of candle soot: "Likewise you must have a care not to lay it too thick, for if it be, you cannot work with that neatnesse and curiosity as otherwise you may" (38). The second deals with the proper whetting of the burin: "It is very necessary that you take great care in the exact whetting of your Graver; for it is impossible that you should ever work with that neatnesse and curiosity as you desire, if your Graver be not very good, and rightly whetted" (44).

52. Boyle, *New Experiments Physico-Mechanicall*, A4r.

53. "Exact" appears nine times; "exacter" two times; "exactly" 24; and "exactnesse" one.

54. Ibid., 11, 13.

55. Faithorne, *Art of Graveing and Etching*, A2r.

56. *The Excellency of the Pen and Pencil, Exemplifying the Uses of them in the most Exquisite and Mysterious Art of Drawings, Etching, Engraving, Limning. Painting in Oyl, Washing of Maps & Pictures* (London, 1668), 58.

57. William Salmon, *Polygraphice; or, The Art of Drawing, Engraving, Etching, Limning, Painting, Washing Varnishing, Colouring and Dying* (London, 1672), 89.

58. Faithorne, *Art of Graveing and Etching*, 44.

59. *Excellency of the Pen and Pencil*, title page.

60. John Evelyn, *Sculptura: or The History, and Art of Chalcography and Engraving in Copper* (London, 1662). 149–50.

61. In the Osbourn Signature Collection at the Beinecke Library, Yale University, there is a page cut from an unknown sale catalogue that shows the title page of *The Art of Graveing and Etching* with John Evelyn's signature on it. Despite great effort on their parts, Moira Fitzgerald and Eva Guggemos of the Beinecke Library were unable to determine the source of the clipping and where Evelyn's copy of *The Art of Graveing and Etching* is now. Their theory is that a private collector bought it.

Evelyn wrote of his discovery of Faithorne's translation in a letter to John Wilkins dated January 29, 1660/61. Evelyn to Wilkins, Early Letters, EL/E/1, Archives of the Royal Society of London.

"Perspicuity and Neatness of Expression"

Algebra Textbooks in the Early American Republic

ROBIN E. RIDER

It requires a *clear* head to be able to see a subject in all its bearings and relations; to distinguish all the niceties and shades of difference between things that bear a strong resemblance, and to separate it from all irrelevant objects that intermingle themselves with it.

George Crabb, *English synonymes* [*sic*] *explained* (1816)

Booksellers in the early American republic offered their clientele a mix of mathematical books: some British works, both old and new, French titles available by special order, and a small but growing number of American works.[1] Their wares included many textbooks, published with the hope of tapping into a growing market in grammar schools, academies, and colleges, and of appealing to students who engaged private tutors and those who endeavored to learn on their own.[2]

The reception, appropriation, and production of mathematical textbooks in late eighteenth- and early nineteenth-century America were informed—and complicated—by competing epistemological and pedagogical traditions, the power of continental mathematical analysis and criticisms of its logical underpinnings, importation of technocratic ideals, institutional rivalries, and commercial competition. To this mix of issues should be added publishing and printing practices in the early Republic—especially choices made in the design and typesetting of mathematics.

This article looks at the influence of well-known and oft-reprinted foreign models—especially textbooks by John Bonnycastle in England and Silvestre-François Lacroix in France—on American teaching of algebra. Different approaches to the learning and value of algebra certainly underlay Bonnycastle's and Lacroix's textbooks.[3] But new and influential typographic practices evident in Lacroix's texts also heightened the contrast. Part of a new style that emphasized open, spare, austere design, with generous white space and smaller type sizes,[4] these typographic practices lent more transparency to the organization of mathematical information. How such typographic practices played out in American mathematics—and in particular in textbooks by Samuel Webber, Jeremiah Day, and John Farrar—in turn speaks to the role of typography in presenting and shaping mathematical learning.[5] Although teaching needs at Harvard and Yale prompted the efforts of Webber, Day, and Farrar, the influence of their textbooks extended well beyond particular institutional contexts.

Typographic Practices in Algebra

The visual culture of mathematics, done well, offers "enormous advantages of *seeing*," as Edward Tufte would say.[6] Readers learn much from the way mathematics is presented in type.[7] Good typography highlights and reinforces ideas; indifferent typography (or worse) obscures ideas and stymies the reader. As the Oxford University Press advised in *The Printing of Mathematics*, "The printer, as a middleman, has the double duty of catching the writer's meaning and of passing it on unimpaired to the reader."[8]

Consider two distinctive typographic approaches in algebra. The first embeds algebraic expressions in sentences (inline or undisplayed). This may require additional vertical spacing (leading)[9] before or after to accommodate exponents or fractions. Algebraic expressions or derivations can also be set off (displayed), either centered or aligned in one way or another, perhaps lined up on the coefficients, or variables, or operational signs (like + and − and =).[10] See figure 5, for example. Most algebra books, then and now, contain a combination of the two approaches; but the use (and design) of displayed algebraic expressions, especially those with careful alignment of signs or similar terms, took on greater significance beginning in the late eighteenth century. It was just at this time that algebra as the study of mathematical structure was being added to algebra as problem-solving and algebra as mathematical language.[11] Hence, particular configurations in print of algebraic concepts during this period were laden with meaning.

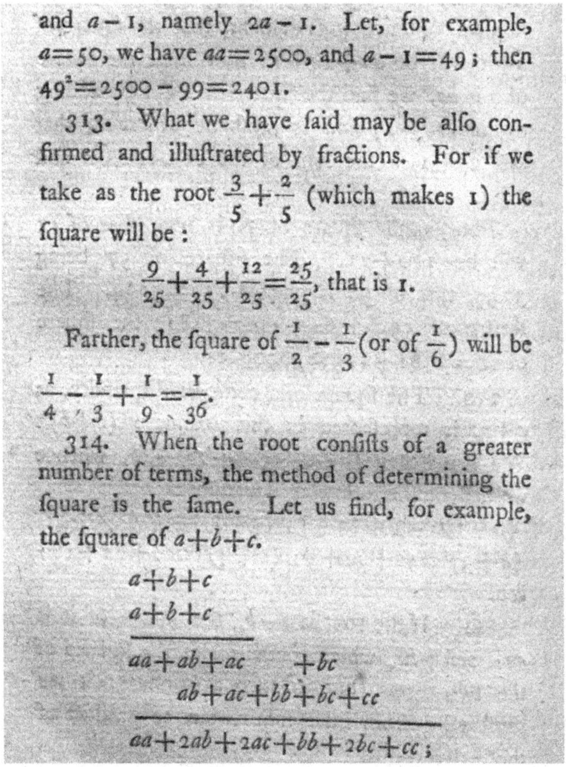

and $a-1$, namely $2a-1$. Let, for example, $a=50$, we have $aa=2500$, and $a-1=49$; then $49^2=2500-99=2401$.

313. What we have faid may be alfo confirmed and illuftrated by fractions. For if we take as the root $\frac{3}{5}+\frac{2}{5}$ (which makes 1) the fquare will be :

$$\frac{9}{25}+\frac{4}{25}+\frac{12}{25}=\frac{25}{25}, \text{ that is } 1.$$

Farther, the fquare of $\frac{1}{2}-\frac{1}{3}$ (or of $\frac{1}{6}$) will be

$$\frac{1}{4}-\frac{1}{3}+\frac{1}{9}=\frac{1}{36}.$$

314. When the root confifts of a greater number of terms, the method of determining the fquare is the fame. Let us find, for example, the fquare of $a+b+c$.

$$a+b+c$$
$$a+b+c$$
$$\overline{aa+ab+ac \qquad +bc}$$
$$ab+ac+bb+bc+cc$$
$$\overline{aa+2ab+2ac+bb+2bc+cc};$$

Figure 5. Inline algebraic expressions; extra leading to accommodate fractions; and displayed expressions aligned, more or less, on operational signs, with blanks left for terms with zero coefficients. From Leonhard Euler, *Elements of algebra* (London, 1797), vol. 1, 154. Euler's influential algebra textbook, first published in 1770, enjoyed multiple editions and translations. (By courtesy of the Department of Special Collections, Memorial Library, University of Wisconsin–Madison)

In the late eighteenth century and the early nineteenth, typesetters[12] and readers, American and otherwise, were well accustomed to rendering, reading, and interpreting structures in print. They regularly dealt with problems of elementary arithmetic (lining up ones, tens, and hundreds with ease), tables, and distinctive, conventional patterns of indentation in poetry.[13] Indeed, students often cut their teeth on arithmetic by copying well-structured examples from a textbook into their exercise books.[14] Printing manuals of the period offered ample guidance about numerals, tables, and poetry, rather less for matters mathematical, except for simple arithmetic.[15]

For anything complicated algebraically, neat and meaningful displays demanded that typesetters develop special expertise and take special pains. The longer and more complicated the algebraic expression, the greater the typographic challenge.[16] Yet careful typesetting, surrounding a mathematical expression with ample white space and aligning similar elements, could offer reinforcement for the author's message, focusing the reader's attention on the structure of that expression, suggesting patterns, and highlighting critical differences. Generous "whites" and extra leading required more paper; careful typesetting of algebraic expressions, complicated by fractions, exponents, nested parentheses, and arcane symbols, demanded more of an expert typesetter's time.[17] And increasing costs of production mattered, especially in an increasingly competitive publishing marketplace. Evidence of such investment in the look of the printed page repays our careful attention.

John Bonnycastle's *Introduction to algebra* (London, 1782)

In this marketplace, the algebraic works of John Bonnycastle (1750?–1821), master and later professor of mathematics at the British Royal Military Academy at Woolwich, proved, and remained, popular.[18] His *Introduction to algebra*, first published in London in 1782, ran to thirteen editions there over four decades. As Bonnycastle himself wrote: "Books of rudiments . . . concisely written, well digested, and methodically arranged, are treasures of inestimable value; and too many attempts cannot be made to render them perfect and complete."[19] The content of his books remained basically the same over decades, treasure or no. Not surprisingly, he drew largely on British models, starting with the *Universal arithmetick* of the immortal Newton.[20] Bonnycastle's books, as produced in London printshops, combined inline notation and displayed expressions—some with terms obviously aligned, others not so. But Bonnycastle, like many other English mathematical authors, tended to privilege prose explanations of algebraic structures rather than letting displayed expressions tell their own stories.[21]

The influence of both Bonnycastle's 1782 text and its typography was felt, years later, as far away as Philadelphia, as is evident from similarities in mise-en-page in Bonnycastle's original presentation of the binomial theorem and in what was advertised as the "first American edition."[22] Across editions, adaptations, and decades, Bonnycastle's *Introduction to algebra* displayed considerable stability in content and form, even to the look of an individual page or line of typeset text; indeed, market forces and publishing practices favored such stability. Publishers saw advantage in reissuing Bonnycastle's *Introduction to algebra*:

not overfilled with complicated algebraic expressions, it was also, hence, relatively easy to reset in type by following the model of a previous edition. Though different print shops necessarily used different stocks of metal type, they nearly always found it easier to mirror line lengths and page breaks of an existing edition than to redesign pages.[23]

Mathematical Books for Harvard:
Samuel Webber's *Mathematics* (Boston, 1801)

One of the American institutions with abiding need for mathematical textbooks was Harvard. Several times in the late eighteenth and early nineteenth centuries it looked within for an author to craft a textbook tailored for Harvard's students rather than adopt someone else's text. Samuel Webber (1760–1810), with two Harvard degrees of his own, had by the beginning of the nineteenth century risen from tutor of mathematics to Hollis Professor of Mathematics and Natural Philosophy at Harvard. Webber's two-volume mathematics textbook, based on more than a dozen years' experience in teaching Harvard students, was first published in 1801. Webber borrowed "from the best authors"—including Bonnycastle—and freely acknowledged his debts: "The parts of the most approved writings, selected for the purpose, are *copied*, with only such alterations, as appeared to be useful" (emphasis mine).[24] The section on algebra, which occupied close to 120 pages in the first volume, mixed general discussion, rules for solution, and examples, and then set problems for students to solve. Some problems had timeless appeal; others spoke more directly to the interests of the young Republic.[25]

Harvard itself was the publisher; the printers, in Boston, were Isaiah Thomas and Ebenezer Andrews.[26] Notation used in Webber's text suggested that Thomas and Andrews did not often have occasion to use specialized mathematical characters: for example, a sideways capital letter "V" made do for "<" and ">."[27] Webber's compilation, like many of his English sources, favored wordy explanations in prose, with algebraic expressions embedded there if necessary. He wrote, for example, of combining "different algebraic quantities": "The business of this operation is to incorporate [them] into one *mass*, or algebraic expression" (emphasis mine).[28] His rather vague "mass" of algebraic quantities could at times come into focus in neat displays, with coefficients, operational signs, variables, and exponents carefully aligned.

Webber's explanations did hint at the significance of arrangement in algebra, pointing to a column "of like quantities" or calling for terms to be "ranged [arranged] according to their dimensions," "according to the powers of some letter in both of them, placing the highest power of it first, and the rest in order." Occasionally Webber also appealed to the student's ability to discern patterns, "rules," and "laws" from a few terms in a series.[29]

But a commitment to consistently clear, well-ordered alignment of terms was lacking, perhaps in part because Webber was incorporating into his course an array of examples borrowed from older textbooks, which likely came with typographic characteristics intact.[30] Even the sturdy regularities of what we know as Pascal's triangle remained hidden in Webber's exposition.[31]

In 1806 Webber became president of Harvard and shortly attended to the production of a second edition of his *Course* in 1808, making only small changes in the text itself.[32] The section on algebra, for example, involved a fairly straightforward resetting of material from the first edition, rendering most mathematical expressions unchanged in appearance. Several years later, Harvard officials thought Webber's textbook might still be of value and authorized negotiations with his widow.[33] Soon, other American institutions of higher learning were also reconsidering their textbook options.

Attention to Arrangement: Jeremiah Day's *Introduction to algebra* (New Haven, 1814)

In this environment, Jeremiah Day (1773–1867), professor of mathematics and natural philosophy at Yale, advanced his *Introduction to algebra* in 1814 as "accommodated to the method of instruction generally adopted in the American colleges."[34] Students, Day argued, deserved a textbook in which the "distinctness of the objects of inquiry, the symmetry of their relations, [and] the luminous nature of the arguments" contributed to their mastery of "a mode of thinking so abstract" as algebra.[35] Like Webber, Day borrowed liberally from other textbooks, looking to French sources as well as to more familiar English models.[36]

Day defined algebra, as had others, as "a general method of investigating the relations of quantities, principally by letters" but emphasized throughout his text the need for "attentive examination" of the parts of quantities. And those parts needed to be arrayed clearly and with attention to their relations: "the several steps of the reasoning, . . . translated into the language of signs and symbols," serve "to place the quantities and their relations distinctly before the eye, and to bring them all into view at once. They are thus more readily compared and understood."[37] This is just one of several passages in which Day addressed directly the apprehension of spatial considerations in typographic form.[38] He observed, "The algebraic mode has often the advantage, not only in being more *concise* than the other, but in exhibiting the *order* of the quantities more distinctly to the eye," directing the student to place "the parts to be compared . . . one under the other" and signaling the importance of such "regular arrangement" as the level of complexity increased.[39]

With an implicit nod to Condillac, Day made the point even more forcefully: "In ordinary language, the numerous relations of the quantities require a series of explanations to make them understood; while, by the algebraic notation, the whole may be placed distinctly before us, at a single view. The disposition of the men on a chess-board, or the situation of the objects in a landscape, may be better comprehended, by a glance of the eye, than by the most laboured description in words."[40]

For example, Day's book deployed simple typographic elements to show *ba* is the same as *ab*: "Let the number 5 be represented by as many points, in a *horizontal* line; and the number 3, by as many points in a *perpendicular* line."[41]

$$\bullet \quad \bullet \quad \bullet \quad \bullet \quad \bullet$$

$$\bullet \quad \bullet \quad \bullet \quad \bullet \quad \bullet$$

$$\bullet \quad \bullet \quad \bullet \quad \bullet \quad \bullet$$

That $3 \times 5 = 5 \times 3$ would be clear at a glance.

Elsewhere in the book, the typesetter honored Day's objective of clear and regular arrangement by leaving room for missing terms and using symmetrical braces [brackets] and other typographic features to highlight similar operations. This style of typesetting was certainly not unique to Day's textbook, but Day's explicit references to the look of an algebraic expression or derivation are worth noting. For example, when Day noted that "a star is sometimes put in the place of the deficient term," he intended the star both to indicate that something was missing and to preserve the structure of the general expression, thus building on the familiar pedagogical function of pattern.[42] And we may conclude that

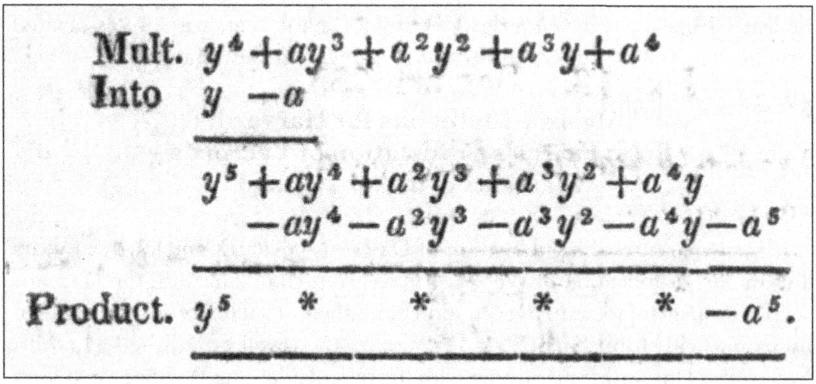

$$
\begin{aligned}
\text{Mult. } & y^4 + ay^3 + a^2y^2 + a^3y + a^4 \\
\text{Into } & y - a \\
\hline
& y^5 + ay^4 + a^2y^3 + a^3y^2 + a^4y \\
& \quad - ay^4 - a^2y^3 - a^3y^2 - a^4y - a^5 \\
\hline
\text{Product. } & y^5 \quad * \quad * \quad * \quad * \quad - a^5.
\end{aligned}
$$

Figure 6. Alignment of operational signs and powers of variables, with asterisks to indicate missing terms. From Jeremiah Day, *An introduction to algebra* (1814), 232. (Digitized by Internet Archive, original from University of California (SRLF), digitized version available at HathiTrust: http://hdl.handle.net/2027/uc2.ark:/13960/t3rv0gs62)

Day's readers took the point, judging from one copy of this edition, in which a contemporary manuscript addition extended the same organizational scheme found in the typeset portion of the chapter.[43]

The example of powers of a binomial often brought out the best in the typesetters of algebra textbooks, and in this Day's book was no exception. The style of presentation, which combined displayed algebraic expressions with prose explanations, required some typesetting finesse in service of a rigorously structured presentation, in which the "regular manner in which these co-efficients are derived one from another, will be readily perceived."[44]

The first edition of Day's *Algebra* enjoyed positive reviews: "very respectable" (if "chiefly compilation"), "most luminous" and "very neat and perspicuous."[45] Though "larger and more expensive" than the texts that had "generally been used," Day's *Algebra* was recommended by one reviewer as likely to "save time, and expense, and trouble"; and Day was to be congratulated "upon the success, which has attended this *peculiar* attempt." But this particular reviewer, citing observations "furnished by a worthy correspondent," also found considerable fault with the printing, which failed to "appear before the publick in so elegant, or so neat, a *dress*," as the content "merits." More to the point, "disadvantages resulting from . . . very bad typography" were particularly troubling in a book "destined to be studied by those, who were, before, totally unacquainted with the science of which it treats." Such criticism points clearly to contemporary expectations and appreciation of type, typography, and presswork.[46]

Day's *Algebra*, with all its virtues and flaws, proved popular, to judge from the many subsequent editions.[47] Some were completely reset in type; stereotyping technology likely produced other nearly identical editions that differed primarily in their title pages.[48] The book stayed among the offerings of the New Haven publisher Hezekiah Howe for decades and for many years was also associated with the name of its printer, Oliver Steele, likewise in New Haven.

Algebra Textbooks for Harvard:
John Farrar's Translation of Lacroix's
Elements of algebra (1818)

In 1815, a year after the publication of Day's *Algebra* and some fourteen years after the first publication of Webber's Harvard textbook in 1801, the Harvard Corporation again took up the question of mathematical textbooks, this time in discussion with John Farrar (1779–1853), by then well established as Hollis Professor at Harvard.[49] Frustrated with the task of adapting Webber's text for a new generation of students, Farrar had thought at one point to substitute Day's *Algebra* (from Yale!) for the algebraic portion of Webber's text, but soon proposed instead compiling an entirely new mathematics course to replace Webber's.[50] By 1818 he was ready with two titles, an introductory algebra text he derived from Euler's elementary treatise on algebra, and, more significantly for our purposes, a "more extended work on Algebra [then to be] followed by other treatises upon the different branches of pure mathematics."[51] The "more extended work" was Farrar's translation of the *Elémens d'algèbre* by Silvestre-François Lacroix (1765–1843), first published in Paris in the revolutionary year VIII for the use of the École centrale des Quatre-Nations in Paris.[52] These translations and adaptations by Farrar, along with the "other treatises upon the different branches of pure mathematics" that would constitute the so-called Cambridge course, demonstrated his close and largely sympathetic reading of the mathematical and pedagogical methods of French mathematicians of the late eighteenth and early nineteenth centuries.[53]

Farrar had undertaken the translation of Lacroix's *Algebra* because it was "approved by the best judges, and . . . generally preferred to the other elementary treatises . . . in France." He took as his text the eleventh French edition of Lacroix's *Algebra* (Paris, 1815), and intended his work "for the use of the students of the university at Cambridge, New England" (i.e., Harvard), though he likely recognized a broader potential market.[54] As with Day's textbook, historians have carefully analyzed the basic content of Farrar's algebra texts, their engagement with critical philosophical questions, and their adoption elsewhere.[55] Rather less attention has been paid to the circumstances of their publication, even less to the typography.

The title page bore the complicated attribution "Printed by Hilliard and Metcalf, At the University Press. Sold by W. Hilliard, Cambridge, and by Cummings & Hilliard" in Boston in 1818. Key in the publishing project was W. (i.e., William) Hilliard (1778–1836), who came from a family of Harvard graduates but had not attended college himself. He instead entered the printing trade and was "engaged as the printer" for a new printing office, called the University Press, established by Harvard.[56] Hilliard's relationship with Harvard was not without strain; but the considerable output of his various printing and publishing ventures clearly benefited from his Harvard ties.[57]

In particular, through his association with science faculty at Harvard, Hilliard saw an American market for solid, well-printed scientific books, and helped to fill it. He would find numerous occasions to work closely with Farrar in a publishing program for mathematics, mechanics, and aspects of natural philosophy; and what became known as the Cambridge series—all of it compiled or translated by Farrar—set important patterns in the look and feel of Hilliard's academic offerings. In particular, Farrar's translation of Lacroix's *Algebra* affords a useful opportunity to explore congruences between algebraic and typographic structures and offers a measure of Hilliard's adaptation of Parisian typographic style to American practice.

Farrar's translation sustained key features of Lacroix's approach, including the advantages of algebraic language, the display of relations among quantities and expressions, and the value of general results. As typeset in Hilliard's shop, it also followed quite closely the typographic layout of the French editions available to Farrar and to Hilliard.[58] Early in the volume, Farrar, following Lacroix, made the explicit connection between algebra and language: "It is by exercising one's self frequently in translating questions from ordinary language into that of algebra, and from algebra into ordinary language, that one becomes acquainted with this science. . . ."[59] In fact, both Lacroix and his American translator displayed clearly the advantages of that algebraic language by means of a large tableau that compared the same problem expressed "by common language" and "by algebraic characters" ("le langage ordinaire" and "l'écriture algébrique"). See figure 7 for Farrar's tableau. In both the American and French versions, the many words in "common language" stood in stark contrast to the economy of algebraic notation, typeset in similar fashion and framed by ample white space.[60] Algebraic language also permitted the reader "to see how the several parts of the quantities concur to form the results."[61]

Farrar faithfully translated Lacroix's precise statement of the numerical obvious—"in numbers the several products present themselves in order, beginning with the units at the last place on the left, on account of the subordination established between the units of each figure of the dividend according to the rank which they hold"—then drew an analogy to the need to impose similar order on the terms in an algebraic expression, according to "the order of the exponents of the power of the same letter, beginning with the highest and proceeding from left to right. . . . This is called *arranging* the proposed quantities."[62] In a numerical array, of course, appropriate blank space was conventionally inserted to align digits in columns of numbers.[63] In algebra, at least in Day's textbook, "A star is sometimes put in the place of the deficient term."[64] In the main, the typesetters both for the French versions and for Farrar's translation of Lacroix's algebra followed the order of exponents from highest to lowest in algebraic expressions, leaving blank spaces—not the asterisks used in Day's textbook—to preserve the alignment of like terms and make obvious the presence or absence of a term of given exponent.[65]

A reader's awareness of symmetry and pattern in mathematical expressions, as aided by careful typographical alignment and parallel arrangement, could even overcome a regrettable typographical error or two. Page 180 of Farrar's translation, for example, featured five substitutions spread across the page: $p/n = P$, $q/n = Q$, and so forth (see figure 8). In the substitution resulting in T, which should have read $t/n = T$, the denominator is either a u or an n upside down, and in at least one copy the t in the numerator seems to be missing altogether. Though such typographical errors appeared in a crucial spot in the derivation, an alert reader, armed with reasonable expectations of symmetry and pattern, would likely soldier on without undue confusion, realizing that t/n was called for.[66]

Like Day's *Algebra*, the Cambridge course was well received. Other American institutions quickly adopted it for use for their own students. A lengthy review offered ample praise for Lacroix's choice of the "most general

SOLUTION.

By common language.	By algebraic characters.
	Let the number to be divided be denoted by a.
	the excess of the middle part above the least by b.
	the excess of the greatest above the middle one by c.
	The least part being x.
The middle part will be the least, plus the excess of the mean above the least.	The middle part will be $x+b$.
The greatest part will be the middle one, plus the excess of the greatest above the middle one. The three parts will together form the number proposed.	The greatest will be $x+b+c$.
Whence the least part, plus the least part, plus the excess of the middle one above the least, plus also the least part, plus the excess of the middle one above the least, plus the excess of the greatest above the middle one, will be equal to the number to be divided.	Whence $x+x+b+x+b+c=a$.

2

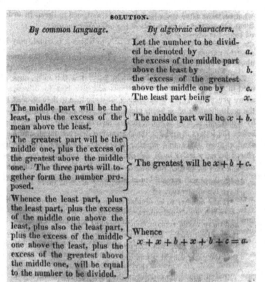

Figure 7. Detail of tableau comparing solutions "By common language" and "By algebraic characters." From S.-F. Lacroix, *Elements of algebra*, trans. Farrar (1818), 9.

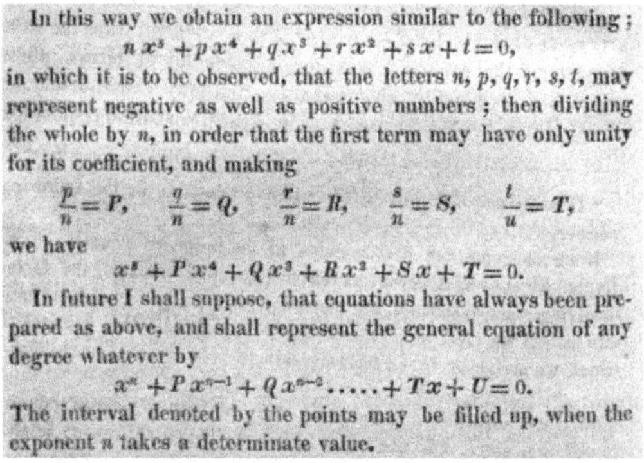

In this way we obtain an expression similar to the following;

$$n x^5 + p x^4 + q x^3 + r x^2 + s x + t = 0,$$

in which it is to be observed, that the letters n, p, q, r, s, t, may represent negative as well as positive numbers; then dividing the whole by n, in order that the first term may have only unity for its coefficient, and making

$$\frac{p}{n}=P, \quad \frac{q}{n}=Q, \quad \frac{r}{n}=R, \quad \frac{s}{n}=S, \quad \frac{t}{u}=T,$$

we have

$$x^5 + P x^4 + Q x^3 + R x^2 + S x + T = 0.$$

In future I shall suppose, that equations have always been prepared as above, and shall represent the general equation of any degree whatever by

$$x^n + P x^{n-1} + Q x^{n-2} \ldots\ldots + T x + U = 0.$$

The interval denoted by the points may be filled up, when the exponent n takes a determinate value.

Figure 8. Typographical error of u for n in the denominator of what should have been $t/n = T$. From Lacroix, *Elements of algebra* (1818), 180. In another copy of this work, held by the New York Public Library, digitized by Google, and available through HathiTrust (http://hdl.handle.net/2027/nyp.33433066405527), the t in the denominator was either barely inked or altogether missing; the u remained the denominator.

methods" and use of "the analytical method," the instrument of invention in mathematical science.[67] Such comments rested on the reviewer's recognition and appreciation of general mathematical structures as highlighted and clarified by clean, even elegant typography.

Reading Algebra
through Day and Farrar

There is no doubt of the influence of Day's and Farrar's algebra textbooks. Sheer numbers of editions over decades and the fact of multiple owners of a given copy suggest that thousands of American students learned their algebra (or at least *some* algebra) from these books.[68]

In Day's textbook they read about a "language of signs and symbols" that could place "quantities and their relations distinctly before the eye, and . . . bring them all into view at once."[69] Day's explanations spoke to the implicit economy of algebraic language, urged the "bare inspection" of the order and arrangement of terms in algebraic expressions, and concluded that regularities and patterns could be "readily perceived." He also wanted students to perceive and appreciate the "beauty of the mathematics." Oddly, Day's prose-heavy approach might have conflicted with his announced objective that algebraic quantities be "more readily compared and understood"; and the typesetting also tended to favor denser pages, especially at the beginning. This embedded "the several steps of the reasoning" with algebraic expressions in the "common mode of writing," rather than set off with leading before and after, even when Day seemed to advocate the opposite.[70]

For Farrar's edition of Lacroix's algebra, by contrast, the typesetter(s) in Hilliard's employ consciously mirrored generously spaced displays of algebraic expressions and derivations found in the French edition of 1815. With "a glance of the eye," a reviewer noted, the reader could "comprehend all the conditions, relations and consequences" of what would otherwise be "complicated and bewildering."[71] Small differences between the Paris and Cambridge versions derived mainly from the fact that the pages set in Hilliard's shop had more lines per page.[72] As Farrar remained faithful, both in translation and in spirit, to Lacroix's focus on "every thing which relates to the composition and decomposition of quantities," so too did Hilliard's shop follow the model of the Parisian typesetters who produced the French edition of Lacroix's *Algebra*.[73] Striking similarities between the typographic configuration of algebraic expressions and derivations in the two versions certainly suggest that the typesetter worked on setting Farrar's translation with a copy of the French edition to hand.[74]

Hilliard's printing and publishing establishment thus supplied several generations of American students, both elementary and advanced, with textbooks informed by Euler's "clearness and elegance of demonstration and illustration," the analytical style of mathematicians of the great schools of revolutionary France as conveyed by Lacroix, and hence access to some of the best continental mathematics of this period.[75] Hilliard's house style for presenting algebra in print, like the French typographic style he emulated, helped not only to commend a brand of mathematics couched in algebraic language, it accustomed its many American readers to clear and distinct displays and thereby enhanced their appreciation for the analytical potential of algebraic structure.

"Perspicuity and Neatness of Expression": Algebra in Print

That which was clear, perspicuous, and neat was held in high esteem in the late eighteenth and early nineteenth centuries.[76] A review in the *Analectic magazine* (1815) of Day's *Treatise of plane trigonometry* cited its "neatness and perspicuity"; another review in the same periodical called for "the utmost perspicuity and brevity" in the expression of "general principles and rules," and commended Day's "clearness of method" and "perspicuity and neatness of expression."[77] A reviewer in the *Southern review*, meanwhile, praised Lacroix's "perspicuous style."[78] Even Farrar, in an otherwise critical review of a potted astronomy text, found its chief merit to be its "plainness and perspicuity."[79]

Existing models from England and France of course influenced the typography of algebra textbooks in the early American republic. Moreover, different typographic styles could be seen as congruent to complementary approaches to the teaching of algebra. Specifically, algebraic expressions embedded in sentences underscored the utility of algebra as a mathematical language, whereas works that featured many well-crafted displays of algebraic expressions served well the engagement of algebra with relations among quantities, patterns, and generality.[80] For example, Farrar, following Lacroix, sought to state "the connexion which exists between the several algebraic operations," this "in order to present this subject in a more clear light."[81]

Not only did typography reflect available models and authorial choices, it also served to make important mathematical points in and of itself. The shape of displayed algebraic expressions—rendered in black type on white paper— could help a reader appreciate and even anticipate symmetries and other mathematical patterns. Set off from the rest of the text, framed with white space, a set of algebraic expressions or succession of equations invited study of its structure and the relations among parts making up the whole.

Historians of the book, describing typography of the late eighteenth and early nineteenth centuries, refer to a typographic esthetic rid of ornament in favor of a neoclassical "purity of form" as expressed on "sparse pages."[82] They might as well use phrases like "neatness and perspicuity" to describe the new typographical style. Whatever the labels, the older examples of Bonnycastle

and Webber, the popularity of Day's introductory algebra text, the mathematical and political affect of Lacroix's textbooks, and the publishing program
of Hilliard and Farrar all speak in one way or another to questions of tradition,
innovation, and influence, both mathematical and typographical. And in these
textbooks we see a complicated interplay of mise-en-page and mathematics, in
which typography as another instrument of thought operated upon algebraic
concepts and thereby shaped both lesson and learning.

Notes

1. For example, in Albany, New York, *The balance and state journal* 1, no. 48 (16 June
1809), 1, as included in the *American Periodicals Series* (hereafter *APS Online*), advertised the
wares of Croswell & Frary's "Book-Store and Printing Office," including "Classical &
School Books, at the lowest prices." The masthead noted "By Croswell & Frary—Printers
to the People." Cummings, Hilliard, & Co. of Boston advertised in their own *United
States literary gazette* 1, no. 6 (1 July 1824), 96 (and also in other issues), that they had
"constantly on hand the most valuable and popular School and Classical Books,"
including those that they had published. The firm could "furnish, on liberal terms, every
book . . . of any value which America affords," and regularly imported books "for a
moderate commission" from England and the Continent. Interestingly, the *Gazette* was
printed in Cambridge "at the [Harvard] University Press, by Hilliard and Metcalf."
Cummings and Hilliard, in partnership, alone, or with others, published numerous
textbooks, including some notable titles on mathematics: see the discussion below of
textbooks for use at Harvard.

2. For example, Consider and John Sterry, *Proposals, for printing by subscription: The
first volume of the American youth, being a system of arithmetic and algebra in two books* . . . (New
London, Conn.: Printed by T. Green, 1788), aimed "to present the Youth of America
with a new and complete course of Mathematical Learning." See Louis C. Karpinski
(with Walter F. Shenton), *Bibliography of Mathematical Works Printed in America through 1850*
(Ann Arbor: University of Michigan Press, 1940), 1; also a table tracking that growth
over time, 15. See "A note on imports and domestic production" signed H.A. (presumably
Hugh Amory) appended to James Raven, "The Importation of Books in the Eighteenth
Century" in *The Colonial Book in the Atlantic World*, vol. 1 in the series A History of the
Book in America, ed. Hugh Amory and David D. Hall ([Worcester, Mass.]: American
Antiquarian Society, 2000), 197–98. As the present essay was going to press, vol. 2 in that
series, *An Extensive Republic: Print, Culture, and Society in the New Nation, 1790–1840*, ed. Robert
A. Gross and Mary Kelley, appeared, with attention both to "Colleges and Print Culture"
(by Dean Grodzins and Leon Jackson) and to "Newspapers and Periodicals" (by Andie
Tucher), along with several contributions on "The Book Trades in the New Nation."

For historical consideration of textbooks, see, for example, *Paradigm: Journal of the
Textbook Colloquium*, at http://faculty.ed.uiuc.edu/westbury/paradigm/. About mathe-
matical teaching and textbooks, see historical studies ranging from Florian Cajori, *The
Teaching and History of Mathematics in the United States* (Washington, D.C.: GPO, 1890), to
Peggy Aldrich Kidwell, Amy Ackerberg-Hastings, and David Roberts, *Tools of American
Mathematics Teaching, 1800–2000* (Baltimore: Johns Hopkins University Press, 2008), esp.

chap. 1 and the works cited there. American textbooks were often issued with a local market in mind, as Karpinski's bibliography makes evident. Although their title pages often promised broad utility, as in Nicholas Pike's *A new and complete system of arithmetic, composed for the use of the citizens of the United States* (Newbury-Port [Mass.]: Printed and sold by John Mycall, [1788]), markets in fact likely proved more local.

3. Specifically, John Bonnycastle, *An introduction to algebra with notes and observations: Designed for the use of schools, and places of public education* (London: Printed for J. Johnson, 1782), and Silvestre-François Lacroix, *Élémens d'algèbre, à l'usage de l'École centrale des Quatre-Nations* (Paris: Duprat, an VIII [1799]), both with numerous subsequent editions.

4. Daniel B. Updike, *Printing Types, Their History, Forms, and Use*, 3rd ed. (Cambridge, Mass: Belknap Press of Harvard University Press, 1962), vol. 2 of 2, 147: "light, open types and widely spaced and leaded pages." Warren Chappell and Robert Bringhurst, *A Short History of the Printed Word*, 2nd ed., revised and updated (Point Roberts, Wash.: Hartley & Marks, 1999), 164, 169, on neoclassicism in typographic design.

5. This examination of significant mathematical textbooks in the first decades after the American Revolution is part of a larger project on ways typographic practices shaped, transmitted, and reinforced mathematical knowledge in the eighteenth and early nineteenth centuries, in which I have drawn on close examination and comparison of multiple editions and translations of mathematical textbooks and printers' manuals, both in print and in digital form; searched reviews and other cultural expressions in nineteenth-century periodicals, especially those in large sets, online or microform; and teased meaning out of bibliographical records.

For other studies of mathematics textbooks of this period, see, for example, Jean Dhombres, "French Mathematical Textbooks from Bézout to Cauchy," *Historia Scientiarum*, no. 28 (1985): 91–137; Kidwell, Ackerberg-Hastings, and Roberts, *Tools of American Mathematics Teaching*; Karen D. Michalowicz and Arthur C. Howard, "Pedagogy in Text: An Analysis of Mathematics Texts from the Nineteenth Century," in *A History of School Mathematics*, ed. George A. M. Stanic and Jeremy Kilpatrick, vol. 1 of 2 (Reston, Va.: National Council of Teachers of Mathematics, 2003), 77–80; Karen Hunger Parshall and David E. Rowe, *The Emergence of the American Mathematical Research Community, 1876–1900: J. J. Sylvester, Felix Klein, and E .H. Moore*, History of Mathematics, 8 (Providence, R.I.: American Mathematical Society, 1994), 2–23; Helena M. Pycior, "British Synthetic vs. French Analytic Styles of Algebra in the Early American Republic," in *The History of Modern Mathematics*, vol. 1 of 2: *Ideas and Their Reception*, ed. David E. Rowe and John McCleary (Boston: Academic Press, 1989), 125–54; Gert Schubring, "On the Methodology of Analysing Historical Textbooks: Lacroix as Textbook Author," *For the Learning of Mathematics* 7, no. 3 (1987): 41–51; Jonathan Topham, "A Textbook Revolution," in *Books and the Sciences in History*, ed. Marina Frasca-Spada and Nick Jardine (New York: Cambridge University Press, 2000), 317–37. My concern here will be more on the "look and feel" of these textbooks and how they transmitted ideas of mathematical structure and less on inductive versus deductive approaches or such issues as the treatment of negative and imaginary numbers.

6. In his *Envisioning Information* (Cheshire, Conn.: Graphics Press, 1990), on 45; see also Tufte's other highly influential books. Michael Twyman uses the phrase "learning through the eye" in "Textbook Design: Chronological Tables and the Use of Typographic Cueing," *Paradigm*, no. 4 (December 1990): [1].

7. And reviewers cared. See the review "Mr. Edward's first *Principles of Algebra*" in *Monthly magazine* [or *British register*] (1 September 1818): 159: "The typography of the work does great credit to the printer."

8. Robin E. Rider, "Early Modern Mathematics in Print," in *Non-Verbal Communication in Science prior to 1900*, ed. Renato G. Mazzolini, Bibliotheca di Nuncius, Studi e testi, 11 (Florence: Olschki, 1993), 91–113, and "Shaping Information: Mathematics, Computing, and Typography," in *Inscribing Science: Scientific Texts and the Materiality of Communication*, ed. Timothy Lenoir (Stanford, Calif.: Stanford University Press, 1998), 39–54; Donald E. Knuth, "Mathematical Typography," American Mathematical Society, *Bulletin*, n.s., 1, no. 2 (1979): 337–72, and his work on TEX; guides like Theodore William Chaundy, P. R. Barrett, and Charles Batey, *The Printing of Mathematics: Aids for Authors and Editors and Rules for Compositors and Readers at the University Press, Oxford*, 3rd impression (London: Oxford University Press, [1965; revised ed. published 1957]), quotation on 22; and William L. Schaaf, "Some Reflections on the Printing of Mathematics," *Mathematics Teacher* (March 1955), 165–66.

9. "Leading," from the pieces of lead used to hold the lines of type apart and in place. See Thomas Curson Hansard, *Typographia . . . With Practical Directions* (London: Printed [by Hansard] for Baldwin, Cradock, and Joy, 1825), 447–52, on "Leads, or Space Lines"; Philip Gaskell, *A New Introduction to Bibliography* (New York: Oxford University Press, 1972), 46. Stanley Morison, *First Principles of Typography* (New York: Macmillan, 1936), 11, notes that "The intelligent use of leading distinguishes the expert from the inexpert printer."

10. See, for example, Chaundy et al., *Printing of Mathematics*, 4, 37, 82, 83, and the dust jacket illustration. Compare with the category of extrinsic "features of graphic language" that "encourage particular reading strategies" in Twyman, "Textbook Design," [2]; also his "The Graphic Presentation of Language," *Information Design Journal* 3, no. 1 (1982): 2–22, esp. 11–16.

11. See Parshall and Rowe, *Emergence*, 2–23; Pycior, "British Synthetic vs. French Analytic Styles"; Robin E. Rider, *A Bibliography of Early Modern Algebra, 1500–1800*, Berkeley Papers in History of Science, 7 (Berkeley: University of California, Office for History of Science and Technology, 1982), among numerous treatments of late eighteenth- and early nineteenth-century algebra. See also Hans Niel Jahnke, "Algebraic Analysis in the Eighteenth Century," in *A History of Analysis*, ed. Jahnke, History of Mathematics, 24 (Providence, R.I.: American Mathematical Society; [London]: London Mathematical Society, 2003), 105–36.

12. Note the distinction between typesetters (compositors), who set type, and the pressmen, who operated the printing presses. Compare with *metteurs en page*, those responsible for design of printed pages (mise-en-page).

13. See William Emerson, *Cyclomathesis: Or, an easy introduction to the several branches of the mathematics. Being principally designed for the instruction of young students, before they enter upon the more abstruse and difficult parts thereof* (London: Printed for J. Nourse, 1763), long division on 84, tables on 126–27. Examples of tables in *The printer's manual, an abridgment of Stower's Grammar* (Boston: R. & C. Crocker, 1817), reproduced as Caleb Stower, *The Printer's Manual* (New York: Garland, 1981), 4, 23, 65. According to John Bidwell's introduction, this "was the first of many American manuals to provide instructions directly derived from British sources" (xiii). Hansard, *Typographia*, appendix, after 920, provided a handsome

two-color example of a table. Compare with Twyman, "Textbook Design," on chronological tables and "structured typography" [6]. Concerning "the modern mode of printing poetry," at least as viewed in late eighteenth century Britain, see M. F., untitled letter to Mr. Urban (dated 8 Feb [1790?]), *Gentleman's magazine* 60, part 2, no. 2 (August 1790): 697–98, citing the benefit of "deep margins" and "lines at a distance" (698). Compare with the anonymous poem "On the art of printing," *The Columbian museum; or, Universal asylum* (Philadelphia: John Parker, 1793), following page 60, which featured indentation of such lines as "Hail Printing! Friend of learning and of virtue!"

14. Kidwell, Ackerberg-Hastings, and Roberts, *Tools of American Mathematics Teaching*, 6, reproduces a page from one such exercise book. For others, see Ashley K. Doar, "Cipher Books in the Southern Historical Collection [University of North Carolina, Chapel Hill]," available in the Carolina Digital Repository, https://crs.lib.unc.edu/, esp. 64. On Abraham Lincoln's cipher book, see Michalowicz and Howard, "Pedagogy in Text," 79–80.

15. For example, Philip Luckombe, *A concise history of the origin and progress of printing with practical instructions to the trade in general compiled from those who have wrote [sic] on this curious art* (London: Printed and sold by W. Adlard and J. Brown, 1770), 253–57 (numerals), 283–86 (braces and fractions), 378 (table), 398–99 (poetry, noting the need for "many thick spaces"), 470–72 ("Of Mathematical, algebraical, and geometrical sorts").

16. Luckombe expressed strong views on the subject: "In Algebraical work, therefore, in particular, gentlemen should be very exact in their copy, and Compositors as careful in following it, that no alterations may ensue after it is composed; since changing and altering work of this nature is more troublesome to a Compositor than can be imagined by one that has not a tolerable knowledge of printing" (ibid., 472).

17. "Hence it is, that very few Compositors are fond of Algebra, and rather chuse [sic] to be employed upon plain work, tho' less profitable to them than the former; because it is disagreeable, and injures the habit of an expeditious Compositor" (ibid., 472). Hansard echoed this opinion, although from his "own practice" found "that the compositor's business may, in this respect, be rendered much more pleasant and profitable by the contrivance of proper materials" (*Typographia*, 453). See also *The London scale of prices for compositors' work . . . 1810* [etc.], 2nd ed. (London: Trade Council of the London Union of Compositors, [1835?]), 72: "Simple algebraic matter is not charged less than double; but when complex in its nature a higher price is paid." *The Printer* (London: Houlston and Stoneman, 1833) advised that "an acquaintance with mathematics and algebra is something of the greatest advantage" for a typesetter: excerpt reproduced as Reading 2.2, "The Compositor," in Richard-Gabriel Rummonds, *Nineteenth-Century Printing Practices and the Iron Handpress with Selected Readings*, vol. 1 of 2 (New Castle, Del.: Oak Knoll Press; London: British Library, 2004), 44.

18. Thomas Whittaker, "Bonnycastle, John (c. 1760–1821)," rev. Adrian Rice, in *Oxford Dictionary of National Biography*, ed. H. C. G. Matthew and Brian Harrison (Oxford: Oxford University Press, 2004), http://www.oxforddnb.com/view/article/2855. WorldCat offers an uncertain birthdate of 1750.

19. Bonnycastle, *An introduction*, [v]. On Bonnycastle's use of the rule method, see Michalowicz and Howard, "Pedagogy in Text," 86.

20. Isaac Newton, *Arithmetica universalis; sive de compositione et resolutione arithmetica liber. Cui accessit Halleiana æquationum radices arithmetice inveniendi methodus* (Cambridge, [Eng.]:

Typis academicis; London: Impensis Benj. Tooke, 1707); translation as *Universal arithmetick: Or, a treatise of arithmetical composition and resolution. To which is added, Dr. Halley's method of finding the roots of æquations arithmetically*, trans. Joseph Raphson, rev. and corr. Samuel Cunn (London: Printed [by T. Wood] for J. Senex [etc.], 1720); and subsequent editions of both.

21. However, for three decades the London publisher for Bonnycastle's algebra was John Johnson, who also published two editions of Francis Horner's translation of Euler's *Elements of algebra* (itself a translation into French from German). These latter editions demonstrated that clear, careful, and ample displays of algebraic expressions were well within the printers' ken.

22. A digitized copy of Bonnycastle, *An introduction* (1782) is available in HathiTrust at http://hdl.handle.net/2027/mdp.39015006973633; likewise, Bonnycastle, *An introduction to algebra; with notes and observations; designed for the use of schools and places of public education*, 1st American ed. (Philadelphia: Joseph Crukshank, 1806), at http://hdl.handle .net/2027/mdp.39015063620382. The "popular passages" feature in Google Books and searches in other full-text compilations turn up even more subtle examples of Bonnycastle's influence on American textbooks. For American editions of Bonnycastle's algebra, see Karpinski, *Bibliography*, 160–62.

23. Reprinting in this period generally required resetting of all the type. Stower, *Printer's manual*, 13. "Stereotyping" would eventually afford relief from this costly practice, but it served to reinforce initial decisions in page design and to sustain typographic convention. See Hansard, *Typographia*, chap. 16. "European literary and philosophical intelligence," *American register; or, general repository of history, politics and science* 2 (2 January 1807), explored advantages and disadvantages of stereotyping and claimed that "the stereotype possesses a security against error that no book from moveable types ever can attain" (¶6 [*APS Online*, 353]).

24. Samuel Webber, *Mathematics: Compiled from the best authors and intended to be the text-book of the course of private lectures on these sciences in the University at Cambridge*, vol. 1 of 2 (Boston: Printed for the University at Cambridge, by [Isaiah] Thomas & Andrews, 1801). The corporation hoped students would derive "great advantages . . . from a judicious work of [this] kind" (Advertisement, [iii]). Digitized copies of both volumes are available in HathiTrust: the former, containing the section on algebra, at http://hdl .handle.net/2027/nyp.33433069078214. Among Webber's sources were English textbooks by Hutton, Bonnycastle, Mole, and Emerson.

25. "A company at a tavern had 81. [£] 15s. to pay for their reckoning; but, before the bill was settled, two of them sneaked off" (Webber, *Mathematics*, 1:368). "At a certain election 375 persons voted, and the candidate chosen had a majority of 91; how many voted for each?" (352).

26. Arrangements for printing and sale of Webber's text suggest the close involvement of the Harvard Corporation in matters of typographic culture. The perceived need for a Harvard printing establishment shortly resulted in corporation approval for buying a press and type. Harvard University Archives (hereafter HUA), Corporation Records, UA I.5.30.2, vol. 4: 1795–1810, 619 (re. 20 April 1802); also 618. See also Max Hall, *Harvard University Press: A History* (Cambridge, Mass: Harvard University Press, 1986), chap. 1. On Harvard's willingness to purchase copies of Webber's mathematics textbook, see HUA, UAI 5.120, Corporation Papers (8 May 1802).

The book's title page read "Boston: Printed for the University at Cambridge, by Thomas & Andrews, 1801." Among other scientific works printed by Thomas and Andrews were titles on geography, surgery, military art and science, and almanacs. See also Thomas's own influential *The history of printing in America, with a biography of printers, and an account of newspapers. To which is prefixed a concise view of the discovery and progress of the art in other parts of the world*, 2 vols. (Worcester, Mass.: From the press of Isaiah Thomas, 1810).

27. William Hilliard, later printer of Harvard textbooks, would also resort to this expedient. Compare with Luckombe, *Concise history*, 470–72; Hansard, *Typographia*, 452–53.

28. Webber, *Mathematics*, 1:272, 271.

29. Ibid., 274, 355, 290. "[B]y obtaining a few of the first terms, the law of the progression will be manifest" (326).

30. Ibid., 276. See, for example, the fairly tidy example 12, although arranging three or more examples across the page (saving paper) complicated the typesetter's task of alignment. In example 7 on page 284, different terms were not segregated into neat columns, though space was sufficient to do so. Returning to more familiar numerical territory, the typesetter aligned without evident difficulty the 1s, 10s, and 100s, as well as variables and coefficients (341, 343).

31. Ibid., 310–11. This was the case despite the invocation of "Sir Isaac Newton's Rule" prescribing the coefficients in integral powers of a binomial.

32. Webber, *Mathematics*, 2nd ed., vol. 1 of 2 (Cambridge: Printed at the University Press by William Hilliard, 1808), advertisements. On the circumstances of publication, see Mary Ann James, "Engineering an Environment for Change: Bigelow, Peirce, and Early Nineteenth-Century Practical Education at Harvard," in *Science at Harvard University: Historical Perspectives*, ed. Clark A. Elliott and Margaret W. Rossiter (Bethlehem, Penn.: Lehigh University Press, 1992), 56, 72–73, esp. the HUA sources cited there; see also Amy K. Ackerberg-Hastings, "Mathematics Is a Gentleman's Art: Analysis and Synthesis in American College Geometry Teaching, 1790–1840" (PhD diss., Iowa State University, 2000), esp. 130n77 and 182n92 about contracting "for the second printing."

33. HUA, Corporation Records, UA I.5.30.2, vol. 5, 1810–1819, 32 (14? August 1811). Discussion evidently continued over several months until a decision to reprint only the arithmetic was made (54–55 [7 February 1812]; 73 [18 May 1812]). As noted in the advertisement in Webber, *A system of arithmetic: Reprinted from the mathematical text-book* (Cambridge: Published and sold by William Hilliard; Hilliard & Metcalf. . . . [*sic*] Printers, 1812), "By the permission of the Corporation of Harvard College, the arithmetical part of Dr. Webber's Mathematics is here published in a separate volume." Ackerberg-Hastings, "Mathematics," describes an "Academic Convention" by 1814 to evaluate textbooks and accomplish "more uniformity between different institutions" (132n81).

34. Jeremiah Day, *An introduction to algebra, being the first part of a course of mathematics, adapted to the method of instruction in the American colleges* (New-Haven: Published by Howe & Deforest; Oliver Steele, Printer, 1814), Preface [3].

35. Ibid., Preface 4–5.

36. He made "free use . . . of the works of Newton, Maclaurin, Saunderson, Simpson, Euler, Emerson, Lacroix, and others," ibid., Preface 6. See discussions in, for example, Ackerberg-Hastings, "Mathematics," chap. 3, esp. 132–34; Stanley M. Guralnick, *Science and the Ante-bellum American College* (Philadelphia: American Philosophical Society, 1975),

chap. 3, esp. 50–52; Kidwell, Ackerberg-Hastings, and Roberts, *Tools of American Mathe-matics Teaching*; Parshall and Rowe, *Emergence*; Pycior, "British Synthetic vs. French Analytic Styles."

37. Day, *Introduction*, 10, 12, 40–41. Similar language on page 28 introduced the addition in algebra of "several distinct quantities," which "must be brought together [and] connected in some form of expression, which will present them at once to our view, and show the relations which they have to each other." Instructions on page 33 specified: "When the quantities to be added contain several terms which are *alike*, and several which are *unlike*, it will be convenient to arrange them in such a manner, that the similar terms may stand one under the other"; and the typesetter complied. On multi-plying a series by a ratio: "To make this plain, let the new series be written under the other, in such a manner, that each term shall be removed one step to the right of that from which it is produced in the line above" (218). That series should be typeset with particular care was not uncommon; what was unusual, and noteworthy, was that Day's exposition emphasized the point.

38. Compare with Twyman on "typographic cueing" and "structured typography" in his "Textbook Design."

39. Day, *Introduction*, 263.

40. Ibid.; on generality as a tool in presenting "relations of quantities . . . to our view," 174. One review of Day's full course, in *Analectic magazine* (June 1817): 441–68, noted Day's prolixity: "He has scarcely left any room for the subject to be rendered plainer by verbal explanation" (468). Rather than chess pieces or objects in a landscape, Day might have adduced other, venerable examples to make the same point. See, for example, John Cruso, *Militarie instructions for the cavallrie: Or, rules and directions for the service of horse: Collected out of divers forrain authors ancient and modern, and rectified and supplied, according to the present practise of the Low-Countrey warres* ([Cambridge, England]: Printed by the printers [i.e., Thomas Buck and Roger Daniel] to the Universitie [*sic*] of Cambridge, 1632), using letters and other typeset characters to demonstrate formations, as on page 47.

On Condillac, *La langue des calculs* (1798), see Robin E. Rider, "Measure of Ideas, Rule of Language: Mathematics and Language in the Eighteenth Century," in *The Quantifying Spirit in the Eighteenth Century*, ed. Tore Frängsmyr, J. L. Heilbron, and Robin E. Rider (Berkeley: University of California Press, 1990), 113–40, esp. 115–20.

41. Day, *Introduction*, 42. Abram Robertson, in "The binomial theorem demon-strated by the principles of multiplication," *Philosophical transactions* 85 (1795): 298–321, had also trotted out a 5-by-3 rectangle divided into 15 square units (301), and used a geometric metaphor to describe the algebraic procedure: "The several products, there-fore, arranged under one another in a perpendicular line" (311, compare with the display on 313). Day referred explicitly to a discussion of binomials in "Phil. Trans. 1795" (presum-ably Robertson's article) in *Introduction*, 233.

42. Day, *Introduction*, 44 (spaces), 46 (braces), 49 (star for "deficient term"). Stars [asterisks] appeared elsewhere in abundance (232). On the utility of symmetry: "The symmetry of these expressions is well calculated [!] to fix them in the memory" (172).

43. See manuscript annotations in a copy of ibid., digitized by Internet Archive, original from University of California (SRLF), available at HathiTrust: http://hdl.handle.net/2027/uc2.ark:/13960/t3rvogs62 (77). Annotations elsewhere in that same copy also show attentiveness to similarity in displays (109).

44. Ibid., 235, with "co-efficients, taken separate from the letters," in triangular array. "The order which these co-efficients observe [display]; is not obvious, like that of the exponents, upon a bare inspection," although as typeset the pattern was in fact clear (236). Further, a one-line equivalent in "algebraic characters" of the wordy binomial theorem (237) demonstrated the economy of algebraic expression.

45. "Domestic literary intelligence," *Analectic magazine* (1 July 1815): 85; "Review," *The literary and philosophical repertory: Embracing discoveries and improvement* (1 January 1816): 336, 332. Day himself had aimed for "the luminous nature of the arguments" (*Introduction*, Preface 4). On the review in the *Analectic magazine*, see Ackerberg-Hastings, "Mathematics," 144–45.

46. "Review," *Literary and philosophical repertory*, 339–40.

47. Pycior, "British Synthetic vs. French Analytic Styles," 148n20. Compare with the assessments in Ackerberg-Hastings, "Mathematics," 145–51, and Parshall and Rowe, *Emergence*, 15–16.

48. See the announcement—"In press and will soon be published, a new edition of the Rev. President Day's System [*sic*] of Algebra, with alterations and additions"—in "Literary and philosophical intelligence," *Christian spectator* (1 July 1819): 375, referring to the edition of 1820.

49. HUA, Corporation Records, UA I.5.30.2. (photocopies), vol. 5, 1810–1819, page 228 (8 May 1815). Compare with the mention of reopening contracts concerning Webber's *Mathematics*, Corporation Papers, UA I.5.120 (27 October 1815). Ackerberg-Hastings, "Mathematics," explores these discussions in the section "John Farrar and Harvard" (161–82). Interestingly, she calls attention to Farrar's correspondence with Day about the issue.

50. Guralnick, *Science and the Ante-bellum American College*, cites Farrar's letter of 22 January to Kirkland, dating it 1817 (54); it may in fact be 1818. Compare with HUA, Tolman index entry for Hilliard & Metcalf, referring to College Papers, vol. 8:46, about an estimate (14 May 1817) for printing an edition of mathematics.

51. Farrar's letter to Kirkland (see note 50). Ackerberg-Hastings, "Mathematics," notes that Farrar "had to publish the translation of Euler's *Introduction to Algebra* at his own expense," this latter based on a letter from the bookseller Francis Nichols to Jeremiah Day (185). The elementary text in question appeared as *An introduction to the elements of algebra, designed for the use of those who are acquainted only with the first principles of arithmetic. Selected* [and translated by John Farrar] *from the Algebra of Euler* (Cambridge [Mass.]: Printed by Hilliard and Metcalf, at the University Press, 1818), Advertisement dated "February, 1818." Farrar's adaptation of Lacroix's textbook on arithmetic would also appear in 1818.

52. Lacroix, *Élémens d'algèbre*. This work was demonstrably popular in France, reaching the eleventh French edition by 1815. For more on Lacroix, see João Caramalho Domingues, *Lacroix and the Calculus* (Basel: Birkhäuser, 2008), including the bibliography; also Dhombres, "French Mathematical Textbooks," on Lacroix as textbook author.

53. Advertisement in Farrar's translation of Euler, *Introduction* (1818). On translation of French mathematical works, see Ivor Grattan-Guinness, "The End of Dominance: The Diffusion of French Mathematics Elsewhere, 1820–1870," in *Mathematics Unbound: The Evolution of an International Mathematical Research Community, 1800–1945*, ed. Karen H. Parshall and Adrian C. Rice, History of Mathematics, 23 (Providence, R.I.: American

Mathematical Society; [London]: London Mathematical Society, 2002), 17–44, esp.
20–24. Grattan-Guinness emphasizes the period after 1820 but does include Lacroix's
Complément des élémens d'algèbre (first edition 1799) and its German translations in 1804 and
1811.

54. S. F. Lacroix, *Elements of algebra. Translated from the French* [by John Farrar] *for the
use of the students of the university at Cambridge, New England* (Cambridge, N.E. [New England]:
University Press, 1818), hereafter cited as Lacroix Farrar Cambridge 1818, advertisement,
dated June 1818.

55. See note 5.

56. Hall, *Harvard University Press*, 6. The second edition of Webber's textbook had
been "Printed at the University Press, by William Hilliard. 1808."

57. Hilliard's name appeared as printer and/or publisher on works from ancient
history to automata, mummies to silkworms, Malthusian theories to midwifery, as well
as on novels, poetry, and many books, pamphlets, and broadsides closely connected
with Harvard, judging from records in WorldCat. See Ackerberg-Hastings, "Mathematics," 167. His relationship with Harvard eventually ended, not without rancor, as
recounted in Hall, *Harvard University Press*, chap. 1.

58. Farrar used as his model Lacroix, *Elémens d'algebre*, 11th ed. (Paris: V[euve].
Courcier, 1815). Careful comparison of other Paris editions dated 1804, 1813, and 1820
indicates their close similarity to the eleventh edition of 1815, although I do note differences when they occur. Subsequent notes use the shorthand of Lacroix Paris 1815,
and so forth.

About the availability of books at Harvard, see, for example, Elliott and Rossiter,
Science at Harvard University, as well as occasional catalogs of Harvard's library. Pycior and
Ackerberg-Hastings, among others, have assessed Farrar's mathematical approach with
regard to analysis and synthesis, French and British influences, algebra and geometry, etc.

59. Lacroix Farrar Cambridge 1818, 12. Compare with references to language on
page 4.

60. In Lacroix Paris 1804, for example, the tableau on page 13, outlined in a double
rule, appeared on a page by itself, with the type set at right angles to the running head.
In Lacroix Farrar Cambridge 1818, Farrar's typesetter arranged it over two pages
in more customary fashion (we would now say "portrait" format, as opposed to "landscape"), though of necessity with narrower columns: "in the following table, . . . against
each step is placed a translation of it into algebraic characters" (9–10). Farrar's translation
preserved Lacroix's view that algebraic notation "abridges very much" the demonstration
of propositions (28). Page 33 also used two columns of a sort, with algebraic expressions
on the right and explanations of each step on the left, similar to Lacroix Paris 1804, 45.
See also the tableau in Lacroix Paris 1815, 13.

Although Hilliard favored "portrait" format on pages 9 and 10, a few pages later he
followed his French models in arraying a complicated problem in perpendicular fashion—
compare Lacroix Farrar Cambridge 1818, 36, and Lacroix Paris 1804, 49. Other works
had likewise accommodated long problems: see, for example, Newton, *Universal arithmetick*,
265; Colin Maclaurin, *A treatise of algebra, in three parts*, 5th ed. (London: Printed for
C. Nourse [etc.], 1788), 38, 206, 240–41. Art. 183 (252 in Lacroix Paris 1804, Paris 1815,
and Paris 1820; 184 in Lacroix Farrar Cambridge 1818) showed the efficacy of an ordered
display compared to the discussion in words.

61. Lacroix Farrar Cambridge 1818, 37. Compare with the same passage in Lacroix Paris 1804, 50: "Les opérations algébriques effectuées sur des quantités littérales, laissant voir comment les diverses parties de ces quantités concourent à la formation des résultats, font souvent connaître des propriétés générales des nombres, indépendamment d'aucun système de numération."

62. Lacroix Farrar Cambridge 1818, 43 (Art. 41), corresponding to Lacroix Paris 1804, 58. Another clear example of order imposed can be found in Lacroix Paris 1804, 249, and Lacroix Farrar Cambridge 1818, 182, where the wording read "Arranging the result with reference to the powers of x."

63. For example, Lacroix Paris 1804, 339; near Art. 245 in Lacroix Paris 1820, 339. These and the foldout tableau in Lacroix Paris 1820, near 261, and the table in Lacroix Farrar Cambridge 1818, 249, afford instructive examples of (slight) difference in the typesetting of mathematical tables. The habit of numerical tables carried over to tidy, well-aligned, instructive arrays of numerical patterns, as in the example, frequently adduced in algebra textbooks, of the powers of 3: compare Lacroix Paris 1804, 40, and Lacroix Farrar Cambridge 1818, 30.

64. Day, *Introduction*, 49. Hansard, *Typographia*, 431, noted the use of the asterisk "to denote an omission, or a hiatus, by loss of original copy, in which case the asterisks are multiplied according to the extent of the chasm."

65. See Lacroix Farrar Cambridge 1818, page 35, where the three reductions were neatly aligned on operational signs, numerical coefficients, variables, and exponents, with spaces (not asterisks) inserted when a digit was not needed. On page 34, however, in an example of multiplication of polynomials in two variables, the alignment broke down fairly completely. As such, it resembled the setting in Lacroix Paris 1804, 46.

66. The French models in Lacroix Paris 1804 and 1820, on page 247 in both editions, did not contain this typographical error.

67. Review of the Cambridge course in *North American review* 4, no. 2 (October 1821): quotations on 363. William Cushing, *Index to the North American Review, vols. 1–125, 1815–1877* (Cambridge: Press of John Wilson and Son, 1878), also reproduced in Kenneth Walter Cameron, *Research Keys to the American Renaissance* (Hartford, Conn.: Transcendental Books, [1967]), identified the reviewer as George Barrell Emerson. Pycior, "British Synthetic vs. French Analytic Styles," describes Emerson as a "tutor of mathematics at Harvard under Farrar" (130). In the discussion of this and other reviews of Farrar's course, Ackerberg-Hastings, "Mathematics," notes Emerson's role in helping to "complete Farrar's translation of Bézout's calculus textbook" (190n120).

68. See, for example, the copy of Day, *Introduction*, digitized and available at http://hdl.handle.net/2027/uc2.ark:/13960/t3rv0gs62, identified in note 43, still in evident use at Yale decades after publication, with the ownership mark Henry C. Leavenworth, Oct. 3, 1845.

69. Day, *Introduction*, 12. Day used similar language on page 28 to introduce the addition in algebra of "several distinct quantities" that "must be brought together. They must be connected in some form of expression, which will present them at once to our view, and show the relations which they have to each other." The instructions on page 33 were explicit: "When the quantities to be added contain several terms which are *alike*, and several which are *unlike*, it will be convenient to arrange them in such a manner, that the similar terms may stand one under the other"; and the typesetter complied with

an example featuring careful alignment of similar terms. The discussion on page 218 of multiplying a given series by a ratio put the point more forcefully: "To make this plain, let the new series be written under the other, in such a manner, that each term shall be removed one step to the right of that from which it is produced in the line above." That examples of series should be typeset with particular care so as to align similar terms was not particularly unusual; what was unusual, and noteworthy, was that Day as author reinforced the message in his exposition.

70. Ibid., 235, 236, Preface 4, 12. The first algebraic expression displayed and centered on the page appeared only on page 19, although a two-column display appeared on page 10.

71. "Review of the Cambridge course of mathematics," *American journal of science and arts* (4 January 1822): 304–26, quotation on 320.

72. Compare, for example, the structured display of the product of four binomials in Lacroix Paris 1815, 252, and Lacroix Farrar Cambridge 1818, 184. I count 10 percent more lines per page, on average. Hilliard's type also appears slightly larger in proportion to the page size than does the type in the French edition.

73. Lacroix Farrar Cambridge 1818, 136–37, corresponding to Lacroix Paris 1815, 187. Indeed, Farrar's rendering of Lacroix's words sometimes called the reader's attention more forcefully to the "disposition of the several parts of the operation" (Lacroix Farrar Cambridge 1818, 135), as compared to the more compact "je dispose l'opération" (Lacroix Paris 1815, 185).

74. This was not unusual in typesetting of this period. It is possible that Farrar tipped his translations of passage after passage into a copy of the French edition or that he cut and pasted the algebraic displays from the French edition into the manuscript that he provided to Hilliard's shop.

75. "Review of the Cambridge course of mathematics" (4 January 1822), 307.

76. The adjectives "neat" and "perspicuous" were applied, for example, to the "type and paper" used in books, as in the advertisement for William Mavor's *Universal history* in James McPherson, *The Poems of Ossian*, with notes by Malcolm Laing, 2 vols. (Edinburgh: Printed . . . for Archibald Constable and Co. [etc.], 1805), and to prose delivered by "elegant" typography, as in the *Eclectic review* (June 1807): 416–78, on 478, about Janson's *The stranger in America*. For another example of praise for the perspicuity of a mathematical work, see Thomas P. Irving, who commended Colson's translation of Agnesi for "the perspicuity with which it is explained" (253) in "For the Port Folio," *The Port Folio* 1, no. 16 (26 April 1806): 252–53. Abram Robertson invoked the virtues of "precision and perspicuity" in "The binomial theorem," on page 314; Farrar (to Kirkland, 22 January 1818) praised Euler "who is as much admired for his perspicuity & elegance as for his depth." Compare with the neat reversal in a review of "An elementary treatise on pleading in civil actions: by Edward Lawes" in the *Monthly anthology, and Boston review* (1 March 1808): 162–65, drawing an analogy between "brevity, perspicuity, and certainty" in legal pleadings and extracting "like an equation in algebra, the real points in controversy, and to refer them, with all possible simplicity, to the court or jury" (163).

77. "Domestic literature and science," *Analectic magazine* 6 (November 1815): 421. The trigonometry volume was available in octavo for $1.25. A short notice of Day's algebra appeared in "Domestic literary intelligence," *Analectic magazine* 6 (July 1815): 85.

See the lengthy review of Day's course, parts I–IV, published in the *Analectic magazine* 9 (June 1817): 441–67, esp. 442, 444. Note the evenhandedly negative comment in the footnote on 456, comparing the downside of the "method rigidly synthetical" of English writers and continental writers "groping their way through the dusky regions of analysis."

78. Review of multiple titles in *The southern review* (1 February 1828): 127.

79. Farrar reviewed David Brewster's work entitled *Ferguson's astronomy* in the *North American review* 6, no. 17 (January 1818): 205–24, quotation on 206. See also Ackerberg-Hastings's comments on this review in her "Mathematics," 170–71.

80. That is, "to see a subject in all its bearings and relations," George Crabb, *English synonymes explained, in alphabetical order; with copious illustrations and examples drawn from the best writers* (London: Printed for Baldwin, Cradock, and Joy and T. Boosey, 1816), s.v. "Perspicuity," 221. An edition was published in Boston in 1819.

81. Lacroix Farrar Cambridge 1818, 243. Lacroix Paris 1815 read: "Pour répandre plus de lumière sur ce sujet, je rappellerai, d'après Euler, la liaison qui existe entre les diverses opérations de l'Algèbre" (332).

82. *Encyclopædia Britannica online*, s.v. "Graphic Design," 2007. See also note 4 above.

The Circulation of Scientific Knowledge in Print

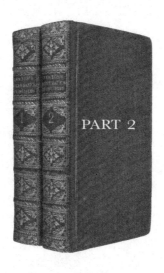

PART 2

Voyaging
and the Scientific Expedition Report,
1800–1940

LYNN K. NYHART

From June to November 1889, the German steamer *National* traversed some sixteen thousand miles of the Atlantic Ocean in a gigantic figure eight, carrying along with its sailing crew a team of six scientists and an artist (see figure 9). Their object was to collect samples of plankton—the tiny organisms that float in the sea and serve as the ultimate food source for most marine creatures—to determine their vertical and horizontal distribution. Led by the Kiel physiologist Victor Hensen, the Plankton Expedition used novel physical and mathematical sampling methods to study the geographical distribution of a new scientific object, the collectivity of floating forms that Hensen had dubbed "plankton." The information gleaned would not only provide new information and insights about life in the ocean but also promised help for the ailing German fisheries industry. If scientists could determine where the plankton were, perhaps they could reliably locate fish as well. With these goals in view, the German government and the various foundations it controlled devoted more funds to this project than had been given to any previous biological study in Germany's history as a unified nation.[1]

To the extent that the history of the Plankton Expedition has been told, it has usually emphasized the controversies over Hensen's self-consciously modern, statistical approach, especially its contrast to the qualitative taxonomic approach of the famous evolutionary morphologist Ernst Haeckel, a specialist in a number of planktonic groups, who bitterly opposed Hensen's enterprise and the funds poured into it. Virtually all the attention given to the publications

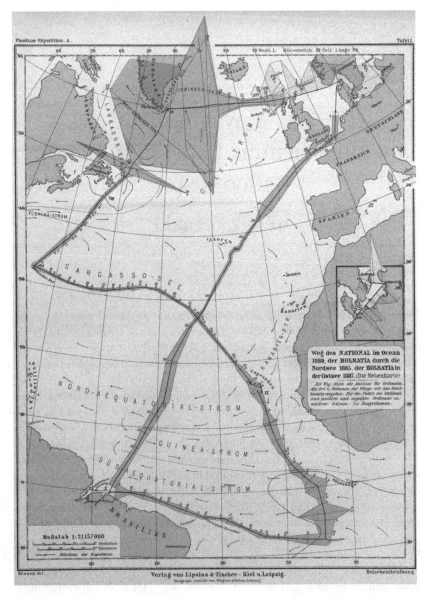

Figure 9. Route of the German Plankton Expedition in the Atlantic Ocean, 1889. From Otto Krümmel, *Reisebeschreibung der Plankton-Expedition*, vol. 1A (1892), plate 1. (Courtesy University of Wisconsin–Madison Libraries)

resulting from the Plankton Expedition has focused on the methodological volume Hensen produced, which became the programmatic foundation for several decades of quantitative plankton studies. But the Plankton Expedition project was much more than simply an expression of Hensen's vision. In fact, the project came to involve thirty-six scientist-authors, ranging from neophytes to prominent zoologists, who produced more than five dozen separate reports under the title *Results of the Plankton Expedition of the Humboldt Foundation*. These *Results* appeared serially over the course of twenty-one years, from 1892 to 1913 (with two laggard reports extending the final publication date to 1926). It was thus a collective project of considerable magnitude and duration, one that spilled over beyond Hensen's own ambitions.[2]

I propose to consider the *Results* from a new perspective, looking beyond the intellectual particularities of Hensen's program to the larger culture of biological knowledge and print production of which it was a part. I argue that multiauthored, serially produced expedition reports such as the *Results* constituted a particular print genre that embodied a prominent form of science in the nineteenth century. What made these reports not merely printed products but significant for our concept of a "print culture" in science is that their production became an integral part of an entire social system of scientific work. Moreover, as this chapter shows, that system evolved over the course of the nineteenth century, shifting in content and growing dramatically in scale. Maintaining the commitment to publish, I would suggest, was in fact what made these projects successful and important as science. (Conversely, the lack of a strong commitment to publish following many voyages often resulted in the collected specimens languishing in boxes for years without ever being analyzed.)[3] The serial trickling out of results, increasingly over decades, meant that the questions asked by the initiators of these projects, and their modes of answering them, were repeatedly placed before the zoological community, thereby shaping the ongoing intellectual substance of zoological science. An understanding of the ways in which scientific knowledge production was enmeshed with the production of the serial expedition report thus deepens our understanding of the character of science in the period.

The history of science and the history of print culture have recently enjoyed considerable mutual support, especially with the growth of interest in the history of popular science in nineteenth-century Britain. James Secord's well-known *Victorian Sensation*, which focused especially on the range of readers for one popular science work, the 1844 *Vestiges of the Natural History of Creation*, is just the best-known tip of the large iceberg. From Jonathan Topham's important early work on the Bridgewater Treatises to Bernard Lightman's recent *Victorian Popularizers of Science*, numerous historians of science and literature have examined popular print as a major locus for science in Victorian culture. The history of popular science writing in other countries has flourished as well, including most significantly Andreas Daum's 1998 *Wissenschaftspopularisierung im 19. Jahrhundert* (Popularization of science in the nineteenth century), which among other things catalogued and analyzed the structure of the publishing market for popular scientific books, periodicals, and book series in nineteenth-century Germany.[4]

As important and exciting as this body of work on popular science is, it by no means fills the analytical domain of the "history of science and print culture." By comparison with studies of popular science, recent analyses of the development of a culture of print within the modern professional scientific community have remained sparse. Yet the possibilities are there. Perhaps the most obvious aspect of the central role played by print culture in the development of science involves scientific journals and journal articles. An older sociological literature treated the history of scientific periodicals and peer review as central aspects of the production of scientific knowledge and community; more recently, interest has been directed at the rhetoric of scientific writing, including scientific article-writing.[5]

Most important scientific knowledge is now expressed through journal articles—indeed, one might claim that for a new bit of scientific knowledge to be recognized as important science, it must appear in a peer-reviewed journal. But the scientific journal is not the only historically significant form of print production within the scientific community. Encyclopedias and dictionaries, textbooks and laboratory manuals, expedition reports and scientific travel narratives, atlases (of geographical places and also of bodies) and histories— each of these can offer insights into aspects of knowledge-making within the scientific community, and opportunities to understand the intersections of print production with knowledge production. Attending to the *Results* and similar expedition reports simultaneously as print productions and as knowledge production can offer new insights into the place of print in the practice of science.

We can profitably examine this form of scientific publication and practice by attending at once to intellectual content, the organization of knowledge production, and communication through print.[6] In the first two sections of this essay, I offer a brief chronological overview of the international tradition of scientific voyaging in the nineteenth century, looking at how the constellation of intellectual aims, organization and scale of work and authorship, and print communication changed over time, as well as some features it retained through those changes. In the third section, I focus in on the Plankton Expedition *Results*, considering the community of people who contributed to them, how they were organized, what they were interested in intellectually, and how the long duration of the *Results* both reflected and shaped intellectual priorities in the German zoological community. In concluding, I speculate on some broader sources of this culture of knowledge- and print-production in nineteenth-century Germany.

The Character of Nineteenth-Century Scientific Voyaging and Publication

Throughout the nineteenth century, the primary context for scientific study on voyages was European imperial expansion. Along with geographical, geological, meteorological, physical, political, and ethnographic information and materials, scientists collected botanical and zoological specimens to contribute to the knowledge of the regions to be conquered, exploited, or otherwise encountered. In this context, the chief aims of natural history inquiry were to discover and classify organisms new to western science and to determine their geographical distribution.[7] Zoological results from this sort of expedition might be published in specialist scientific journals or in the house journal of a sponsoring museum but could also appear within a set of "scientific results" or "reports" published in book form, often comprising multiple volumes.

The high-end standard for the "scientific results" form appears to have been set in early-nineteenth-century Paris, by two expedition-results publishing projects. The *Description of Egypt* (*Description de l'Égypte*), which the French state published between 1809 and 1822, following Napoleon's military and imperial campaign in Egypt from 1798 to 1801, drew together the work of more than 150 scholars, and hundreds more artists and engravers, to present encyclopedic descriptions and images of Egypt's antiquities, modern state, and natural history. The *Description* comprised nine books of text, ten books of plates (plus one devoted to describing the plates), as well as two further elephant folios of plates (measuring 1 meter by .8 meter) and an atlas of maps the same size. The French government poured huge resources into this publication. About ten scientists and engineers contributed to the two volumes of natural history, which included analyses of geology, meteorology, and botanical and zoological specimens and were accompanied by two volumes of plates.[8]

The other significant exemplar was not a state venture but a private one. The thirty-volume *Voyage de Humboldt et Bonpland* (Voyage of Humboldt and Bonpland) appeared between 1805 and 1834, with the parts also published separately. It combined the analysis of specimens (especially botanical specimens) collected on Alexander von Humboldt and Aimé de Bonpland's voyage to South America with philosophical and historical essays, geographical results, atlases, and a travel narrative. This was a different kind of operation from one involving an entire government corps of scholars, but it did involve multiple researcher-authors. Although Humboldt and Bonpland are credited as the overall authors, in fact quite a number of other scholars also contributed to this project, including the zoologists Georges Cuvier and Pierre André Latreille and the botanist Karl Sigismund Kunth. Like the *Description of Egypt*, it also entailed a sumptuous outlay of resources, especially for the production of the plates: Humboldt spent his entire personal fortune on it. The *Voyage* of

Humboldt and Bonpland appears to have been uniquely grandiose in concep-
tion and production for a private natural history publication venture in the
period; nevertheless, it probably served as an aspirational model for later voyage
publications.[9]

Three features are notable about these two French publication ventures.
First, the "scientific results" were distinct from narratives of voyages and exploration, which had a much longer and deeper tradition. The latter generally incorporated natural history observations into a travel account organized by the
course of the voyage, though essaylike observations and reflections were often
interpolated.[10] Indeed, Humboldt's famous three-volume *Personal Narrative* of
his voyage—which comprised volumes 28–30 of the *Voyage* collection just
discussed—fell into exactly this category.[11] In contrast to voyage narratives,
"scientific results" typically eschewed a narrative form, instead presenting data
and its analysis. With respect to natural history, a second feature was that specimens were farmed out to specialists for analysis. In the cases of both Humboldt
and the *Description of Egypt*, those specialists could be found in Paris, usually with
connections to official institutions such as the Museum of Natural History
and the Institut de France (especially its natural science arm, the Academy of
Sciences). Here we can see patterns of social organization and publication
already in place that would serve as a standard for natural history expeditions
and their published scientific results for the rest of the century. Finally, the publication process itself represented commitments of long duration by individual
editors and publishers—in the case of the Humboldt-Bonpland work, nearly
thirty years—and, for the *Description of Egypt*, a long-term commitment by the
state.[12]

The British Empire was larger than the French, especially following the
defeat of Napoleon in 1814–15. But consonant with the less professionalized
status of science in Britain, the British government's approach to scientific
voyaging and publication of scientific results was more modest. Here the *Beagle*
voyage on which Charles Darwin famously served presents a useful exemplar.
Although the nearly five-year-long voyage of HMS *Beagle*, December 1831 to
October 1836, is remembered largely for its inspiration of Darwin's theory of
evolution by natural selection and for his popular travel account (his *Journal of
Researches*, known most often as *The Voyage of the Beagle*), its primary purpose was
imperial: hydrographic surveying to locate safe routes for British military and
commercial vessels. It also resulted in a set of natural history publications,
produced separately from the voyage narrative, under Darwin's aegis. In addition to three volumes on the geology of the voyage that Darwin wrote up as a
single author, he also edited a collective publication, *The Zoology of the Voyage of
H.M.S. Beagle* (1838–43).[13] This was a five-part publication comprising nineteen
numbers, totaling 632 large quarto pages and 166 plates. It involved five British
authors besides Darwin. Darwin distributed his materials according to area of

expertise; the volumes' authors reported separately on fossil mammals (Richard Owen), living mammals (George R. Waterhouse), birds (John Gould), fish (Leonard Jenyns), and reptiles (Thomas Bell). Under Darwin's firm hand, all the parts were completed within seven years of the voyage's return.[14]

The size and scope of the *Beagle Zoology* is on a par with some other important British voyage reports of the period, such as John Richardson's four-volume *Fauna Boreali-Americana* (1829–37), which supplied a precedent for Darwin's request to the British Crown for a supplementary sum of £1000 to support the cost of producing plates.[15] Although the *Zoology* of the *Beagle* was much less lavish and involved than Humboldt and Bonpland's *Voyage* volumes, it has been considered "sumptuous" by Darwin's leading book historian, suggesting a difference in scale of both science and high-end publication practices between France and Britain in the first half of the nineteenth century.[16] Despite these differences, throughout the century there would persist a general pattern of collective publication of "scientific results," which connected imperial voyaging, classificatory analysis along with descriptions of habits and geographical distribution, and considerable attention to the most expensive part of the process: the production of illustrative plates. However, in the latter part of the century, a modification of this pattern would emerge alongside it, as the ocean itself became an object of scientific investigation.

Exploring the Oceans

In contrast to voyages of surveying or exploration during which some marine collecting was done, sustained deep ocean exploration as an object in itself was a new phenomenon of the second half of the nineteenth century.[17] The British HMS *Challenger* voyage around the world (1872–76) set a new benchmark for the scale of marine research, and not only in Britain. Indeed, both Darwin's and Richardson's edited collections of scientific results, substantial as they may have seemed for Britain in the 1830s and 1840s, were vanishingly small in comparison to the output of the *Challenger* voyage. The *Report on the Scientific Results of the Voyage of H.M.S. Challenger* may have been titled in the singular, but it involved seventy-six authors—mainly from Britain, but also from Germany, Belgium, the United States, the Netherlands, and Russia—who produced reports ranging from sixteen pages long to over a thousand, ultimately comprising fifty tomes and some thirty thousand pages. Forty of these volumes were devoted to zoology, and most of these to marine invertebrates. Commensurate with the project's great size and complexity, it took a considerable time to be published: fifteen years, from 1880 to 1895.[18]

The *Challenger*'s scale of research and publication production would itself be taken as a challenge by other nations, including Germany. Having unified as a nation only in 1870–71, Germany was a latecomer to the game of imperial competition, including scientific ocean voyaging (though not, as we shall see in the final section, to the general structure and mode of large collective publishing projects). Major German government-sponsored scientific voyages began with the round-the-world hydrographic survey of SMS *Gazelle* in 1874–76 and would continue up to World War II.[19]

The Plankton Expedition reports, with three dozen authors and approximately sixty-five folio reports, were the first to approach the *Challenger*'s in the scale of production.[20] This would be followed by similarly large-scale marine science expeditions and publications in the later 1890s and early 1900s. The timescale for publication also stretched out: despite Hensen's optimistic estimate that his project would be completed within three years, it continued in earnest for twenty-four years after the specimens were initially collected. Stretching even longer, the German Deep-Sea Expedition of 1898–99, with results authored by some two dozen scientists, produced its final, thirty-seventh report in 1940.[21] If one basic part of the structure remained the same from Darwin's voyage to the Plankton Expedition and after—the distribution of specimens to specialists who analyzed them and then returned their reports to the editor for publication—the dramatic increase in number of authors and duration made the later reports a phenomenon of a different order.[22]

What motivated this new research on ocean life and the expansion of scale that attended it? When Darwin published the *Zoology of the Beagle*, the prime object of zoological research was classification (especially of vertebrates and insects), with attention to life histories and the distribution of animals, which generally informed classificatory decisions. But in the 1840s, marine invertebrates became hot new objects of research. By the end of that decade, every zoologist, it seemed, was working on these creatures, examining their strange forms, life cycles, and taxonomy. (Darwin got in the act, too, publishing four volumes on living and fossil barnacles between 1851 and 1855.)[23] Marine biology, understood as morphology and classification, would continue to dominate zoological research through the 1850s and 1860s.

By the time of the *Challenger* voyage, new intellectual challenges and opportunities for exploring the ocean had arisen. First, Darwin's theory of evolution had intervened. Evolution gave new impetus to marine zoology because of the view that life had originated in the sea. The task of reconstructing the tree of life, then, with hopes of finding representatives of ancestral forms in present-day primitive organisms, necessarily made the ocean an important place.[24] Additionally, Darwin and Wallace's evolutionary theory, and Wallace's later work on zoogeography, helped make questions of distribution central to the unraveling of evolutionary relationships, and much zoological work (especially that undertaken in museums) would focus on questions of distribution.

Deep-ocean exploration added literally a new dimension to zoogeographical questions (a vertical one), as zoologists argued over whether life could exist on the deep-ocean floor; whether there existed an "intermediate fauna" in the realms below the surface depths where light could reach and above the sea floor; and if so, what its characteristic organisms and their distribution would look like. The *Challenger* expedition confirmed that life did indeed exist in the deep ocean, but scientists remained uncertain about the other questions, and these would continue to motivate marine exploration to the turn of the twentieth century. Such questions not only were tied to aspects of evolutionary theory but also involved the exploration of the physical limits of life itself. Especially in the context of marine research, these questions would be closely allied to what we today would call autecological ones, concerning the physical and biological aspects of the ocean environment—temperature, pressure, water chemistry (especially salinity), light (or in the deep ocean, its absence)—that conditioned the distribution of life in the ocean.[25] This would be the main direction of research pursued by German marine scientists from the 1870s forward.

In Germany, as elsewhere, the study of classification and geographical distribution did not close out the reasons for studying the ocean. At least as important for German state interest in the ocean in the 1880s and 1890s was the more recent rise of an ecological-economic perspective aimed at improving fisheries. Indeed, the Plankton Expedition grew directly out of these slightly different interests.

In the late 1860s, after Prussia acquired Schleswig-Holstein (thus substantially expanding Germany's coastline on the North and Baltic Seas), the Prussian government established the Commission for Investigating the German Seas, based at the University of Kiel and spearheaded by Kiel's physiology professor Victor Hensen and several colleagues. Through the 1870s and 1880s, the Kiel commission sought correlations between the distribution of marine organisms and physical attributes of the ocean such as temperature and salinity, as well as biological correlations among organisms. Hensen was particularly interested in the role of plankton in what he called "the metabolism of the sea," which he saw as key to understanding the distribution of fishes. The Plankton Expedition and its follow-up research were aimed at establishing the quantity and distribution of plankton. Hensen's method proved controversial in part because it depended on extrapolating from samples to make generalizations about distribution—a method foreign to most biologists at the time. But if Hensen's approach was idiosyncratic in its unique focus on plankton and its quantitative analysis, his attention to the vertical and horizontal distribution of organisms participated in a broader trend toward an increasingly ecological orientation in marine biogeography.[26]

Darwinian evolution, marine biogeography, and marine ecology were three important intellectual developments that gave direction and impetus to marine biology as a major subfield of zoology in the later nineteenth century. The Plankton Expedition was an important instantiation of these developments, and its post-voyage work exemplified how research might be organized to pursue them. It is therefore to this work and its representation in the *Results* of the Plankton Expedition that we now turn.

The Plankton Expedition's *Results* and the Organization of Research

The *Results* were organized at the outset into five volumes (though they were not all bound as such by the publisher). Volume 1 appeared as three separate fascicles, devoted respectively to a voyage narrative, geophysical observations (both by the voyage's physical oceanographer, Otto Krümmel), and a methodological treatise by Hensen. Volume 5, also by Hensen, was an overview of the results; it appeared in 1911, before all the individual reports had been returned. Volumes 2 through 4 comprised the analyses of individual taxonomic groups and were printed in over five dozen fascicles, each appearing separately upon its completion. A running analytical table of contents of the whole project, printed on the cover of each fascicle, preserved the conceptual ordering of the project (into taxonomic groups), while underlining of topics and authors indicated those reports that had actually appeared. In this way it is possible to trace, from successive front covers, the minor shifts in classification, organization, and authorship of sections from what was projected to what was produced over time. The detailed nature of the *Results* also allows us to reconstruct how the work of the project was organized and the vital role played by the projected print product in structuring that work. At the same time, inspection of individual fascicles shows how some authors moved beyond Hensen's aims to develop their own.

The first fascicle to be published, volume 1A, was primarily a voyage narrative written by Otto Krümmel. This volume played into a number of the expectations established for such narratives. Following the route of the voyage in his account, he described in vivid detail how the plankton collecting was done and conveyed the repetition involved in sampling over a hundred stations, through fine and stormy weather, through the ice floes below Greenland and the fog off Newfoundland to Bermuda, back across the Atlantic through the rich aquatic life of the Sargasso Sea to the Cape Verde Islands, heading further southeast to Ascension, and then turning west to Para on the northern coast of Brazil before heading in a long northeast shot to England and then home. Its copious illustrations of voyage scenes, sketched by the accompanying artist Richard Eschke and engraved by Heinrich Riffarth, took advantage of nineteenth-century developments in printing technology to make the narrative friendly to a general audience (see figure 10). It was not published separately but appeared as the first volume of the *Results,* in the same large format (33.5 cm by 27.5 cm) as the rest of the volumes—a format better suited to the presentation of statistical tables and maps (which were abundant in some of the later

Ein Schnabelwal wird gefunden.   57

Kommandobrücke aus in der Ferne steuerbord voraus einen grösseren weissen Gegenstand im
Wasser treiben. Indem wir darauf zu hielten, wurde schnell festgestellt, dass es der Körper
eines todten auf der Seite liegenden Walfisches war, was durch Schaaren von Sturmvögeln
(Fulmarus glacialis), die auf und neben ihm Platz genommen hatten und unsere Annäherung
kaum beachteten, bestätigt wurde. Da die ruhige See es ermöglichte, wurde das kleine Schiffs-
boot ausgesetzt, und der Kapitän mit den Zoologen und dem Marinemaler liess sich näher
heranrudern, um den Kadaver ans Schiff zu bugsiren. (Fig. 7). Zwei Sturmvögel in grauem
Jugendkleide wurden dabei zunächst das Opfer unvorsichtiger Gefrässigkeit. Mit einem Boots-
anker wurde der Kadaver gefasst und lag alsbald längsseit des Schiffes. Das ganze Thier zu
konserviren, würde zu viel Aufenthalt verursacht haben, es wurde seine Länge mit dem Bandmaß
zu 8,2 m gemessen und
dann mit vieler Mühe
in mehrstündiger Arbeit
der Kopf abgetrennt.
Der Versuch, den Kör-
per mit Hilfe der Dampf-
winde umzuwenden, um
zu untersuchen, ob nicht
etwa eine Harpune an
der anderen Seite sass,
misslang, die Gummi-
streifen des Akkumu-
lators zerrissen, bevor
das Sicherheitstau des
letzteren wirken konnte,   (Fig. 7.) Der Walfischkadaver wird ans Schiff bugsirt.
woraus hervorging, dass

Figure 10. Illustration embedded in Krümmel's travel narrative. The caption reads,
"The whale cadaver is towed to the ship." Drawing by Richard Eschke, photogravure
by H. Riffarth. From Otto Krümmel, *Reisebeschreibung der Plankton-Expedition*, vol. 1A
(1892), 57. (Courtesy University of Wisconsin–Madison Libraries)

volumes) than to a basic fireside read. Moreover, its narrative chapters were
interspersed with chapters of preliminary results by scientists on the voyage.
Both of these attributes made the overall volume much more self-consciously
scientific than many earlier such narratives—it was part of the project, not the
advertisement for it.

The second part of volume 1 in the organizational schema (though not in
fact the second fascicle to appear) was Hensen's, on the research methodology
of the project. With its detailed description of his sampling and counting
methodology, this fascicle—1B—became the Ur-text for future quantitative
approaches to marine ecology. Because of the controversies over Hensen's
method, historians have often discussed this volume, but they have not treated
in any detail the larger publication project of which it was a part.[27]

What, then, of the rest of the reports, the ones that comprised volumes 2
through 4? These fascicles represent the specialists' analyses of the organisms
found in the samples, undertaken by zoologists across northern Europe. Just as

previous expedition leaders had distributed organisms by rough classificatory groups to specialists in those classificatory areas, so too did the Plankton Expedition. In fact, for the quantitative analysis to work, there needed to be qualitative analysis as well, for "plankton" was a functional category, not a taxonomic one. Referring to all small drifting creatures in the ocean, it included a huge variety of forms: multicellular and unicellular organisms, eggs and larvae of both verte- brates and invertebrates, organisms that spent their whole lives floating in the sea, and stages of organisms that in other parts of their life cycle were sessile. To learn if there were patterns in the distribution of different taxonomic groups, specialists needed to identify which ones were represented in the plankton hauls at different locations. This is the work that made up the bulk of the reports, took so much time, and, I contend, offers us the most insight into the project's role in the zoology community, with respect both to the authors and the content of their science.

The authors represented a spectrum of scientific experience. Some came from the core cadre of young project scientists who worked at the University of Kiel under the supervision of Hensen, the professor of physiology, and Karl Brandt, the professor of zoology. (Unusually, the government grants that sup- ported the project itself funded a half-dozen two-year positions for counters and sorters of the material.) Carl Apstein, author of three reports, ran the day- to-day work at the lab from the return of the voyage forward; Hans Lohmann, author of another three reports, gained his university teaching credential, the *Habilitation*, with his analysis of groups from the Plankton Expedition, and would stay connected to Kiel for the rest of his career. Johannes Reibisch began his analysis of pelagic marine worms (Phyllodocidae) at Kiel, where he worked for the project in 1892–93; part of the research published in the *Results* would serve as his dissertation under the Leipzig zoologist Rudolf Leuckart.[28]

As the project continued, new people, often students, moved into the mix. In Brandt's institute, the diverse protozoan phylum of radiolaria were especially popular. Arthur Popofsky began working on the radiolarian class Acantharea in the fall of 1902, and the first part of his analysis served as his dissertation in 1904. Similarly, Ferdinand Immermann was a student of Brandt when he worked through the expedition's Aulacanthidae (another radiolarian group) in 1904. Many of the other zoologists, outside of Kiel, were likewise younger workers still establishing their reputations—Otto Maas, for example, began his analysis of the craspedote medusae in the Berlin zoology institute in the spring of 1890 when he was twenty-three, spent some time comparing these to materials in Naples, and then returned to Berlin to complete the project in the fall of 1892. As these and the previous examples suggest, Plankton Expedition reports could double as formal academic qualifying publications, and many of these younger workers would continue to pursue marine zoology for their careers. For example, Adolf Borgert, contributor of eleven fascicles, began his work in 1893, three years after completing his PhD at Kiel, and would work on specimens of the Tripyleen family of radiolarians off and on for the next thirty years, as he rose through the ranks to professor at the University of Bonn. Arthur Popofsky, who produced fascicles in 1904 and 1906, also contributed the last fascicle to appear (in 1926). In his spare time as an educational administrator (Oberstudiendirektor) in Magdeburg, he would analyze radiolarians sent to him from various later marine expeditions.[29] Indeed, one impact of the Plankton Expedition was to create a cadre of experts in marine zoology, who contributed to the continuing prominence of this area within German zoology well into the twentieth century.

Other authors were already more established. In addition to Brandt, who contributed one fascicle, these included Karl Chun, who took on the ctenophores (comb jellyfish) and siphonophores (jellyfish-like colonial invertebrates) and had been a professor at Königsberg since 1883; he would later head up the German Deep-Sea Expedition. Georg Pfeffer (who produced a major report on the cephalopods, the group containing squids) had been at the Hamburg zoology museum since 1880. A few experts from beyond Germany lent their expertise on particular taxonomic groups to the project: the prominent Belgian zoologist Édouard van Beneden, for example, took on the Anthozoa (the class that includes corals and sea anemones).[30]

Nearly all the contributors were men, reflecting the lack of higher educational opportunities for women in Germany at the time. The only women named as authors were Marianne Plehn, working in the zoology institute of her mentor Arnold Lang at the university in Zurich (the first of the German-speaking universities to admit women); and Maria Grosset Dahl, a well-educated Russian émigrée who had been hired early on as a sorter and illustrator for the project in Kiel and who would become the wife of one of the original planktonologists on the expedition, Friedrich Dahl.[31] Maria Dahl's contribution speaks to the poignant situation of the rare German female scientist at the time. In the foreword to her analysis of the copepod genus *Corycaeinen*, she apologized twice over. How was it justifiable, she asked, that someone without a PhD in zoology could conduct a taxonomic analysis on creatures as complex as the copepods? She answered in her own defense that she had assisted Brandt in the Kiel Zoology Institute for seven years. Her second apology was for how long it had taken to complete the task: "Since I could work on this only during my spare hours, and I had to dedicate most of my daytime to my four children, even the work of the preliminary sorting took several years to complete."[32]

The contributions' content was nearly as wide-ranging as their authors' experiences. The scientists involved in the detailed research on these planktonic forms did generally count their specimens and worked to identify horizontal and vertical distribution patterns for their groups—tasks critical to Hensen's own project. But the actual reports also show them using those same specimens

to make original contributions to zoological research. Lohmann speculated on
the role of water pressure and other physical factors in shaping the Halacarina
(water mites); in 1906, Paulus Schiemenz, head of the new Leipzig Institute for
Inland Fisheries, discussed the developmental stages of the Pteropods (sea-snails)
and their relationship to phylogenetic development. Valentin Haecker reorganized the preliminary classification and sorting of polychaete (worm) larvae
undertaken by Apstein. Ludwig Rhumbler, known as a contributor to a physico-
mechanical account of development, sought to establish a new classification of
the Foraminifera based on "selectionist and mechanical-physiological factors,"
arguing that hard structures would beat out soft ones in the struggle for life and
provide the engine for reconstructing phylogenetic relationships. Karl Chun
even expressed skepticism toward Hensen's quantitative approach.[33] These
scientists did not even remotely confine themselves to the quantitative results of
plankton distribution. As these reports show, Hensen and his colleagues running
the project did not exercise tight control over the kinds of information produced
by the experts. The scientists beyond Kiel (and to some extent, there as well)
acted as independent scholars and not merely technicians for Hensen's project.
The system allowed multiple aims to be achieved.

The open, distributed nature of this work may well have contributed to the
long duration of the *Results'* serial production, with significant consequences.
From 1892 through 1913, anywhere from one to six fascicles appeared annually,
excepting only a short hiatus in 1902 and 1903, when no reports appeared. The
project stopped producing reports altogether with the outbreak of World War
I; thereafter, only two more reports appeared: one in 1922, and the last in 1926.
A few projected installments were never published. As the published reports
trickled out, institutes and libraries would put them on their shelves and review
journals would mention them.[34] In this way, the long duration of the publishing
project helped to keep these marine organisms, and the distributional and
ecological questions common to all the reports, before the larger community for
decades, serving first to increase attention to them and ultimately to normalize
them as topics within the zoological community.

The Plankton Expedition contributed to an entire system of voyaging and
large-scale, installment-structured report production that, from the 1870s until
World War I, was dedicated to questions of marine-zoological classification,
distribution, and ecology. This included major voyages and expeditions (and
their results) such as the German Deep-Sea Voyage of 1898–99, as well as the
German South Pole expedition from 1901 to 1903, which, though primarily
geophysical and geographical in design, devoted eleven of its twenty volumes
to zoology (and most of these to marine invertebrates). The system might also
be seen to include smaller expeditions that produced significant serial out-
puts such as the Hamburg Museum's expedition to the Straits of Magellan in
1892–93 (the *Results* of which appeared in fifty fascicles between 1896 and

1907), and a privately funded expedition to Spitzbergen in 1898 that produced the *Fauna Arctica* (1900–1933), both of which, though including studies of insects and vertebrates among their volumes, were heavily dominated by marine invertebrates.[35]

This systematic aspect, and the role of serial publication in it, is critical for our understanding of the intellectual development of marine zoology in this period. Different voyages and reports overlapped with one another, such that the problem orientation of new voyages was established in response to reports from previous ones; in addition, because the reports took so long to come out, writers within the same project might be responding to results from other parts, produced earlier. Thus a major theme running from the *Challenger* voyage onward was the question of "bipolarity"—how to explain the morphological similarities of marine organisms found at the poles but separated by temperate oceans in between.[36] Scientists working on different organisms with different assumptions about the nature of distribution, evolution, earth history, and adaptation participated in a long and contentious dispute over this problem. It motivated the Hamburg expedition to the Straits of Magellan, and Pfeffer, who worked at the Hamburg museum, brought it up in his 1912 report on the cephalopods of the Plankton Expedition. To take an overlapping case, the German Deep-Sea Expedition, led by Chun, began producing its results in 1902; Chun himself analyzed the cephalopods collected from that expedition between 1906 and 1910. Pfeffer, still grinding away at his Plankton Expedition report on the cephalopods, folded a comparison with Chun's results into his 1912 report.[37] Ferdinand Immermann's 1904 report on the Aulacanthids, a family of radiolaria, responded to Haeckel's report on that group from the *Challenger* expedition, and also to Adolf Borgert's analysis of closely related Tripyleen families from the Plankton Expedition itself. As these examples suggest, even though the projects were not literally open-ended, they came to possess one of the attributes of journals and other serials: the production of ongoing published conversations and debates among scientists.[38] The expedition-report system as a whole thus served as an engine of scientific research that helped reinforce and advance the zoological community's focus on marine biology.

The Scientist as Volunteer and the Scholarly Culture of Print

As I have indicated, a central feature of the late nineteenth-century large-scale science represented by the expedition report was its distributed organization: it was produced by a geographically dispersed group of scholars, each of whom analyzed particular taxonomic groups of specimens and then sent their analysis to the project editor for publication. This was as true for Darwin's coauthors as it was for Hensen's some fifty years later. In concluding, it is worth briefly

considering more broadly what values and attributes of science this organization embodied. This form of organization had deep roots in the cultural system
of knowledge production in the nineteenth century, not only in natural history,
and not even only in the natural sciences. At its heart, this system was what
made the expedition report an instantiation of a *culture* of print in nineteenth-
and early-twentieth-century science, and not just a genre of scientific print.

One feature of this system that has been little remarked upon is that the
authors of report installments were rarely paid for them. Unlike popular science
writers, who were often paid by the word (indeed, that is often why they wrote
popular science), these experts were expected to donate their time to their
science. I do not mean that they went unpaid for being scientists. To be sure, a
gentleman like Darwin had no need of a salary to support his research, and his
friend Leonard Jenyns earned his living as a clergyman. By contrast, the museum
men Gould and Owen, to whom Darwin farmed out his specimens, did get
paid for their labors as museum curators, and Bell was a professor. Late-
nineteenth-century German zoologists, too, were usually paid to be part of a
university or museum. But these scientists were generally not paid to analyze a
particular set of specimens from a particular voyage—they voluntarily chose to
undertake this investigation, perhaps instead of a different one. This was part
of the great value system of freedom of inquiry associated with science. Where
did this system of organization, which was central to the production of reports
and so fully embodied the ideals of voluntarism and cooperation, come from?

There were a number of roots here. These include the long tradition associated with natural history expeditions (long before they became focused on
marine invertebrates), which, as we have seen, grew over the course of the
century. One must also recognize the publishing tradition of the large, encyclopedic product to which many hands contributed, which included encyclopedias
and dictionary projects from Buffon's *Histoire Naturelle* to the *Oxford English
Dictionary*. Here I would also call our attention to a different kind of project
especially prominent in Germany, which had many of the same properties as
the one embodied in the ocean expedition report system: something we might
call the classical heritage project.

In nineteenth-century Germany, the biggest scientific publication projects were not in the natural sciences at all, but in history, classics, and language
study—all sciences in the German sense of *Wissenschaften*. The *Monumenta Germaniae historica*, a project of collecting and editing sources for German history,
was probably the largest and longest-term of these. Organized in 1819, its first
volumes appeared in 1826, and it continues to produce volumes today. Similarly,
the *Corpus Inscriptionum Latinarum*, created in Berlin in 1847 to pull together all
existing ancient Latin inscriptions into a scholarly collection, began producing
volumes in 1853 and continues to do so today: it currently stands at seventeen "volumes" (each running to multiple tomes) with seventy parts.[39] Such

projects—which in the nineteenth century involved scholarly volunteers collecting materials (often in the field, in the case of the Latin inscriptions), analyzing data, and publishing editions organized in a somewhat taxonomic style—would appear to resemble nothing if not a big natural history project. But in Germany, at least, the analogy more properly goes the other way around: the large-scale kind of work involved, with many scattered scholars contributing to a centralized task that may have no end, seems to have been accepted as normal for the historical-philosophical sciences at mid-century, when this large scale and duration of production was still a rarity in natural history publications. In the German-speaking lands, classical scholarship had much higher academic prestige than the natural sciences in the early nineteenth century and even into the middle part of the century. Perhaps zoologists saw this as a good model—indeed, one that might provide work for generations of zoologists and perhaps bring zoology the same kind of prestige previously held by classical and historical scholarship.

There is much more to be explored here—the role of scientific academies in orchestrating many of these projects, for example, begs for more attention, as does the role of publishers; and it would be worthwhile to compare these projects to large-scale cooperative ventures that studied geophysical phenomena such as magnetism and meteorology.[40] As I have suggested, the classical heritage project was certainly not the sole model of this approach for late nineteenth-century zoologists, but it should not be ignored as a model just because it does not lie within what we consider science today. The large-scale, long-duration, distributed serial report project is likely the offspring of numerous progenitors. If we want to understand the culture of print in science, we need to attend to science's place in the broader scholarly culture of print.

Notes

1. Ernst Haeckel made the claim that this was the largest amount spent to that date for biological research. David M. Damkaer and Tenge Mrozek-Dahl, "The Plankton-Expedition and the Copepod Studies of Friedrich and Maria Dahl," in *Oceanography: The Past*, ed. Mary Sears and Daniel Merriman (New York: Springer, 1980), 465.

2. Victor Hensen, *Methodik der Untersuchungen*, vol. IB (1899) of *Ergebnisse der in dem Atlantischen Ocean von Mitte Juli bis Anfang November 1889 ausgeführten Plankton Expedition der Humboldt-Stiftung*, ed. Victor Hensen (Kiel: Lipsius & Tischer, 1892–[1926]), hereafter *Ergebnisse*. In addition to Damkaer and Mrozek-Dahl, "The Plankton-Expedition," which focuses mainly on the Dahls, see John Lussenhop, "Victor Hensen and the Development of Sampling Methods in Ecology," *Journal of the History of Biology* 7, no. 2 (1974):319–37; Eric Mills, *Biological Oceanography: An Early History, 1870–1960* (Ithaca, N.Y.: Cornell University Press, 1989), esp. chap. 1; Rüdiger Porep, *Der Physiologe und Planktonforscher Victor Hensen (1835–1924): Sein Leben und sein Werk* (Neumünster: Karl Wachholtz

Verlag, 1970), esp. 103–20; and Olaf Breidbach, "Über die Geburtswehen einer quantifizierenden Ökologie—der Streit um die Kieler Planktonexpedition von 1889," *Berichte zur Wissenschaftsgeschichte* 13, no. 2 (1990):101–14.

3. This happened, for example, with many of Louis Agassiz's collections from his South America trip in the 1860s and with collections from a number of American Arctic exploring expeditions in the nineteenth century. See Mary P. Winsor, *Reading the Shape of Nature: Comparative Zoology at the Agassiz Museum* (Chicago: University of Chicago Press, 1991); Michael F. Robinson, *The Coldest Crucible: Arctic Exploration and American Culture* (Chicago: University of Chicago Press, 2006).

4. James A. Secord, *Victorian Sensation: The Extraordinary Publication, Reception, and Secret Authorship of Vestiges of the Natural History of Creation* (Chicago: University of Chicago Press, 2000); Jonathan R. Topham, "Beyond the 'Common Context': The Production and Reading of the Bridgewater Treatises," *Isis* 89, no. 2 (1998): 233–62; Bernard Lightman, *Victorian Popularizers of Science: Designing Nature for New Audiences* (Chicago: University of Chicago Press, 2007). See also G. N. Cantor and Sally Shuttleworth, eds., *Science Serialized: Representation of the Sciences in Nineteenth-Century Periodicals* (Cambridge, Mass.: MIT Press, 2004), as well as the excellent website *Science in the Nineteenth-Century Periodical*, http://www.sciper.org/; Faidra Papanelopoulou, Agustí Nieto-Galan, and Enrique Perdriguero, eds., *Popularizing Science and Technology in the European Periphery, 1800–2000* (Burlington, Vt.: Ashgate, 2009); Andreas W. Daum, *Wissenschaftspopularisierung im 19. Jahrhundert: bürgerliche Kultur, naturwissenschaftliche Bildung und die deutsche Öffentlichkeit, 1848–1914* (München: R. Oldenbourg, 1998), esp. parts 5 and 6; on book series, esp. 324–30.

5. For a résumé of older literature on peer review and scientific periodicals, see the opening paragraphs and citations in Lynn K. Nyhart, "Writing Zoologically: The *Zeitschrift für wissenschaftliche Zoologie* and the Zoological Community in Late Nineteenth-Century Germany," in *The Literary Structure of Scientific Argument*, ed. Peter R. Dear (Philadelphia: University of Pennsylvania Press, 1991). On the rhetoric of scientific writing, see the other articles in the same volume, as well as Charles Bazerman, *Shaping Written Knowledge: The Genre and Activity of the Experimental Article in Science* (Madison: University of Wisconsin Press, 1988), and Greg Myers, *Writing Biology: Texts in the Social Construction of Scientific Knowledge* (Madison: University of Wisconsin Press, 1990).

6. The active role of publishers in this circuit of knowledge production is another important topic that unfortunately cannot be addressed in any detail in this essay.

7. Janet Browne, *The Secular Ark: Studies in the History of Biogeography* (New Haven: Yale University Press, 1983); idem, "British Biogeography before Darwin," *Revue d'Histoire des Sciences* 45, no. 4 (1992): 453–75.

8. Commission des sciences et arts d'Égypte [de la France], *Description de l'Égypte, ou Recueil des observations et des recherches qui ont été faites en Égypte pendant l'expédition de l'armée française* (Paris: Imprimerie impériale, 1809–21). Here the publishing tradition followed was surely that of the eighteenth-century *Encyclopédie*, known for both its comprehensive articles and its lavish technical plates. See Denis Diderot, *A Diderot Pictorial Encyclopedia of Trades and Industry*, ed. Charles C. Gillispie (New York: Dover, 1959), especially Gillispie's introduction.

9. For the complex publication history of Humboldt and Bonpland's *Voyage*, which involved multiple publishers and editions, see Julius Löwenberg, "Alexander von

Humboldt. Bibliographische Uebersicht seiner Werke, Schriften und zerstreuten Abhandlungen," in *Alexander von Humboldt: Eine wissenschaftliche Biographie*, ed. Karl Bruns (Leipzig: Brockhaus, 1872), 2:485–552, esp. 496–521. Assessment of the uniqueness of this publication comes from my own survey of travel literature compiled from WorldCat, using combinations of search terms such as "scientific results," "results," "voyage," "expedition," and equivalents in French and German.

   10. Astonishingly, there appears to be no general historical overview of the travel narrative. For examples of analyses of varieties of early modern and Enlightenment travel narratives in relation to science, see Daniel Carey, "Compiling Nature's History: Travellers and Travel Narratives in the Early Royal Society," *Annals of Science* 54, no. 3 (May 1997): 269–92; Miles Ogborn and Charles W. J. Withers, "Trade, Travel, and Empire: Knowing Other Places 1660–1800," in *A Concise Companion to the Restoration and Eighteenth Century*, ed. Cynthia Wall (Malden, Mass.: Blackwell, 2005); Florence Hsia, *Sojourners in a Strange Land: Jesuits and Their Scientific Missions in Late Imperial China* (Chicago: University of Chicago Press, 2009). On Romantic travel writing, see especially Nigel Leask, *Curiosity and the Aesthetics of Travel Writing, 1770–1840: "From an Antique Land"* (Oxford: Oxford University Press, 2004).

   11. Alexander von Humboldt, *Relation historique du Voyage aux régions équinoxiales du Nouveau Continent . . .*, 3 vols. (Paris: Schoell, Maze, Smith & Gide, 1814–25). The title is rendered in its standard English translation as "Personal Narrative of Travels to the Equinoctial Regions of the New-Continent, during the years 1799–1804."

   12. Humboldt's main French publisher was the firm F. Schoell in Paris, which published many but not all of the volumes of the *Voyage*; he also worked with a half-dozen other publishers on different volumes of the work. See Löwenberg, "Alexander von Humboldt," 498.

   13. For the history of publications resulting from the *Beagle* voyage, see especially the introductions to individual volumes of *The Complete Works of Charles Darwin Online*, http://darwin-online.org.uk/contents.html#books.

   14. Charles Darwin, ed., *The Zoology of the Voyage of H.M.S. Beagle, during the Years 1832–1836*, in 5 parts (London: Henry Colburn, 1838–43). Different copies may be found bound variously in two, three, or more volumes, indicating they were most likely bound by the purchaser.

   15. For the details of the British government's contributions to financing the publication, see Frederick Burkhardt and Sydney Smith, eds., *The Correspondence of Charles Darwin*, vol. 2, *1837–1843* (Cambridge: Cambridge University Press, 1987), especially the introduction and letters 373 and 373a. John Richardson et al., *Fauna Boreali-Americana; or, The Zoology of the Northern Parts of British America . . .* (London: John Murray, 1829–37).

   16. See R. B. Freeman's introduction to *The Zoology of the Voyage of H.M.S. Beagle* in *The Works of Charles Darwin: An Annotated Bibliographical Handlist*, 2nd edition (Folkstone, Kent, England: Dawson, 1977), 26, accessed via the authoritative *Complete Works of Charles Darwin Online*, http://darwin-online.org.uk/EditorialIntroductions/Freeman_Zoology OfBeagle.html. By contrast, Humboldt chose Paris for the production of his work because publishers there were already capable of the lavish publication he had in mind. (See Löwenberg, "Alexander von Humboldt," 497). The larger scale of French voyage reports would persist: see, e.g., Francis Castelnau, *Expédition das les Parties centrales de l'Amérique du Sud . . .*, 17 vols. in 15 (Paris: P. Bertrand, 1850–59).

17. Helen Rozwadowski, *Fathoming the Ocean: The Discovery and Exploration of the Deep Sea* (Cambridge, Mass.: Belknap Press of Harvard University Press, 2005).

18. C. Wyville Thompson and John Murray, eds., *Report on the Scientific Results of the Voyage of H.M.S. Challenger during the Years 1873–76*, 6 vols. in 50 (Edinburgh: Printed for H. M. Stationery office [by Neill and Company], 1880–95). Harold Burstyn, "'Big Science' in Victorian Britain: The *Challenger* Expedition (1872–76) and Its *Report* (1881–95)," in *Understanding the Oceans: A Century of Ocean Exploration*, ed. Margaret Deacon, Tony Rice, and Colin Summerhayes (London; New York: UCL Press, 2001), 49–55. Burstyn claims that the cost of the *Challenger* voyage and its *Report* between 1872 and 1899 was "one of the largest in history" before World War II (51).

19. The *Gazelle* expedition ultimately produced a five-volume collection of expedition reports, but the major zoological results had already appeared in specialist journals; the volume on the zoology and geology of the voyage, prepared by the voyage's zoologist Theodor Studer, instead concentrated on the conditions of existence controlling the distribution of the organisms in the ocean. Hydrographisches Amt des Reichs-Marine-Amts, ed., *Die Forschungsreise S.M.S. "Gazelle" in den Jahren 1874 bis 1876*, vol. 3, *Zoologie und Geologie* (Berlin: E.S. Mittler & Sohn, 1889).

20. The compilation of authors of the Plankton Expedition results is my own, from direct examination of the fascicles of the *Ergebnisse* at the University of Wisconsin–Madison Biology Rare Book Collection, the incomplete online version at the Biodiversity Heritage Library, and entries in WorldCat. I have been unable to find a complete published listing of all the authors and fascicles.

21. Carl Chun, ed., *Wissenschaftliche Ergebnisse der Deutschen Tiefsee-Expedition auf dem Dampfer "Valdivia" 1898–1899*, 24 vols. in 37 (Jena: G. Fischer, 1902–40).

22. Here one wishes for more information about the *Results'* publisher, Lipsius & Tischer, in Kiel and Leipzig. Judging by its publications from the late nineteenth century, when the project began it was a small publisher catering primarily to the university community in Kiel. Beginning with the Plankton Expedition, however, it became a significant publisher of marine biological, ecological, and oceanographic works. If the duration of its commitment to the Plankton Expedition came as a surprise to the publisher, it nevertheless appears to have reaped some benefit from it, making a name in this area. A key aspect of the financial arrangement may have been the decision not to price the entire project ahead of time but to price each volume separately as it was produced. Subscribers (expected to be libraries, scientific institutes, and perhaps a few individual experts), who committed purchasing the entire project, received a 10 percent discount on each fascicle as it came out. (See the publisher's explanation inside the cover of each fascicle.) More research into the financial and leadership role of publishers in such works is a clear desideratum, since they are so obviously central to perpetuating the print culture of science.

23. The first of these was Charles Darwin, *A Monograph on the Sub-Class Cirripedia: With Figures of All the Species* (London: Ray Society, 1851). For the publishing history of this and the subsequent volume on living barnacles, plus the two volumes on fossil barnacles, see the excellent, detailed discussion in the sections "Living Cirrepedia" and "Fossil Cirrepedia" under "Books" in *The Complete Works of Charles Darwin Online*, http://darwin_online.org.uk/contents.html#books.

24. This was the basic premise of Anton Dohrn, who founded the world-famous Naples Zoological Station to pursue this research program. See e.g., Christiane Groeben and Irmgard Müller, *The Naples Zoological Station at the Time of Anton Dohrn*, trans. Richard and Christl Ivell (Naples, Italy: Naples Zoological Station, 1973); Keith R. Benson, "Review Paper: The Naples Stazione Zoologica and Its Impact on the Emergence of American Marine Biology," *Journal of the History of Biology* 21, no. 2 (1988): 331–41, esp. 338.

25. On the role of ocean exploration in increasing attention to ecological and geographical issues in zoology in the decades around the Plankton Expedition, see Lynn K. Nyhart, *Modern Nature: The Rise of the Biological Perspective in Germany* (Chicago: University of Chicago Press, 2009), esp. 323–40. For a broader survey, see Eric L. Mills, "Problems of Deep-Sea Biology: An Historical Perspective," in *The Sea*, ed. Gilbert T. Rowe, vol. 8, *Deep-Sea Biology* (New York: John Wiley and Sons, 1983), 1–79.

26. On the Kiel commission, see Eric Mills, *Biological Oceanography*, chap. 1. On Möbius's and Hensen's collaboration, see Reinhard Kölmel, "Zwischen Universalismus und Empirie—die Begründung der modernen Ökologie- und Biozönose-Konzeption durch Karl Möbius," *Mitteilungen aus dem Zoologischen Museum der Universität Kiel* 1, no. 7 (1981): 17–34, esp. 24–26. More generally on changes in the linkages of life-cycle studies to classification and ecology, see Lynn K. Nyhart, "Natural History and the New Biology," in *Cultures of Natural History*, ed. Nicholas Jardine, James Secord, and Emma Spary (Cambridge: Cambridge University Press, 1996), 426–43.

27. Victor Hensen, *Methodik der Untersuchungen, Ergebnisse*, vol. IB (1899); John Lussenhop, "Victor Hensen"; Olaf Breidbach, "Über die Geburtswehen."

28. On the organization of the project at Kiel, see Eric Mills, *Biological Oceanography*, 27; this work also discusses Apstein and Lohmann, esp. 176–83, 133–38, and sources cited therein; on Reibisch, see Reibisch, *Ergebnisse* 2.H.c: *Die pelagische Phyllodociden* (1895), 3; Reibisch, *Die pelagischen Phyllodociden der Plankton-Expedition der Humboldt-Stiftung*, Inaugural-Dissertation, University of Leipzig (Halle a. S.: Karras, 1894); and Klaus Wunderlich, *Rudolf Leuckart: Weg und Werk* (Jena: VEB Gustav Fischer Verlag, 1978), 47.

29. See the following fascicles of the Plankton Expedition *Ergebnisse*: Popofsky, 3.L.f: *Acantharia Teil I: Acanthometra* (1904), 3; Immermann, 3.L.h.1: *Aulacanthiden* (1904); Maas, 2.K.c: *Craspedote Medusen* (1893), 3; Borgert, 2.E.a.C: *Doliolen* (1894, his first contribution); Borgert, 3.L.h.2: *Bau und Fortpflanzung der Tripyleen: Allgemeiner Teil* (1922; his last contribution); Popofsky, 3.L.h.13: *Coelodendridae (einschl. Coelographidae Haeckel)* (1926).

30. Brandt, *Ergebnisse*, 3.L.a: *Tintinnodeen* (1906–7); Chun, 2.K.a: *Ctenophoren* (1898), 2.K.b: *Siphonophoren* (1898); Pfeffer, 2.F.a: *Cephalopoden* (1912); van Beneden, 2.K.e: *Anthozoen* (1898). On Pfeffer, see Herbert Weidner, *Bilder aus der Geschichte des Zoologischen Museums der Universität Hamburg* (Hamburg: Selbstverlag des Zoologischen Instituts und Zoologischen Museums der Universität Hamburg, 1993), esp. 190–91.

31. Plehn, *Ergebnisse*, 2.H.f: *Polycladen* (1896), 3; M. Dahl, 2.G.f.1: *Corycaeinen* (1912); on Dahl, see Dankaer and Mrozek-Dahl, "The Plankton-Expedition," esp. 469.

32. Maria Dahl, *Ergebnisse*, 2.G.f .1: *Corycaeinen* (1912), 1.

33. Lohmann *Ergebnisse*, 2.G.a.ß: *Halacarinen* (1893); Schiemenz, 2.F.b: *Pteropoden* (1906); Haecker, 2.H.d: *Die pelagischen Polychaeten- und Achaetenlarven* (1898); Rhumbler, 3.L.c. *Foraminiferen: Zweiter Teil* (1911), 3; Chun, 2.K.b: *Siphonophoren* (1897), 7.

34. The results were regularly summarized in the standard German review journals such as the *Zoologischer Jahresbericht* and the *Geographische Jahrbücher* and cited in the specialist literature.

35. Naturhistorisches Museum in Hamburg, *Ergebnisse der Hamburger Magalhaensischen Sammelreise 1892/93*, 50 parts in 3 vols. (Hamburg: L. Friederichsen, 1896–1907); Fritz Römer et al., *Fauna arctica: Eine Zusammenstellung der arktischen Tierformen, mit besonderer Berücksichtigung des Spitzbergen-Gebietes auf Grund der Ergebnisse der Deutschen Expedition in das Nördliche Eismeer im Jahre 1898*, 6 vols. (Jena: G. Fischer, 1900–1933). These German projects themselves may be seen as elements of a larger international system of marine and polar exploration, involving both competition and cooperation, too large to be examined here.

36. This topic has yet to receive its historian, though it motivated much research, including the 1892 Hamburg trip to the Straits of Magellan.

37. Pfeffer, *Ergebnisse*, 2.F.a: *Cephalopoden* (1912), 784, 796, 797.

38. On the journal form as providing open-ended conversation, see Thomas Broman, "Periodical Literature," in *Books and the Sciences in History*, ed. Marina Frasca-Spada and Nick Jardine (Cambridge: Cambridge University Press, 2000), 225–38, esp. 226.

39. On the history of the *Monumenta*, see Harry Bresslau, *Geschichte der Monumenta Germaniae historica* (Hannover, Germany: Hahnsche Buchhandlung, 1976 [reprint of 1921 original]), and Susan A. Crane, *Collecting and Historical Consciousness in Early Nineteenth-Century Germany* (Ithaca, N.Y.: Cornell University Press, 2000). On the history of the Corpus Inscriptionum Latinarum, see the useful information at the *CIL*'s website: http://cil.bbaw.de/cil_en/dateien/forschung.html, including the booklet by Manfred G. Schmidt, *Corpus Inscriptionum Latinarum* (Berlin: Berlin-Brandenburgische Akademie der Wissenschaften, 2001).

40. The *CIL* and the *Monumenta Germaniae* were both run through the Prussian Academy of Sciences in the late nineteenth century, along with such other scholarly publishing projects as the *Deutsches Wörterbuch*. One of the main sources of funding for the Plankton Expedition was the Humboldt Stiftung, which was also controlled by the Prussian Academy. (See the *Sitzungsberichte* of the Königlich Preussischen Akademie der Wissenschaften zu Berlin for annual reports on the different projects and endowments.) Renato Mazzolini has provocatively suggested that national history and language projects provided academies with a new raison d'être in the nineteenth century: "Nationale Wissenschaftsakademien im Europa des 19. Jahrhunderts," in *Nationale Grenzen und internationaler Austausch: Studien zum Kultur- und Wissenschaftstransfer in Europa*, ed. Lothar Jordan and Bernd Kortländer (Tübingen: Max Niemeyer Verlag, 1995), 245–60. On international cooperation in the study of geomagnetism (but without reference to publications), see John Cawood, "Terrestrial Magnetism and the Development of International Collaboration in the Early Nineteenth Century," *Annals of Science* 34, no. 6 (1977): 551–87. On weather data collection from non-scientist volunteers, see Katherine Anderson, *Predicting the Weather: Victorians and the Science of Meteorology* (Chicago: University of Chicago Press, 2005), esp. 99–105.

Crossing Borders

The Smithsonian Institution and Nineteenth-Century Diffusion of Scientific Information between the United States and Canada

BERTRUM H. MACDONALD

In late spring 1876, a letter written by Roderick MacFarlane, an employee of the Hudson's Bay Company stationed in Fort Chipewyan, Athabasca District (now northern Alberta), reached Spencer Baird, assistant secretary of the Smithsonian Institution. "I have been sadly disappointed for the last year," MacFarlane wrote, "not a single Harper publication has come to hand for 1875, and none so far for 1876. I'd sooner have gone without many other things. I wrote you on this subject last winter. I regret to be giving you so much trouble year after year; but you are so very kind & obliging that I am tempted to trespass on your good nature. Let me know, please, the cost of the volumes & papers & how we stand. My $50.00 may be assumed with what I lately directed to be sent for the [buying?] purposes, while any balance owing I will only be too glad to remit later on."[1] Living far from metropolitan centers, MacFarlane keenly felt his isolation from the latest developments in science when publications did not arrive regularly. That he relied on the Smithsonian Institution as his supplier of scientific publications was not unusual.

As the body of scientific literature grew substantially in the nineteenth century, the Smithsonian Institution, established in Washington, D.C., in 1846 from the bequest of the Englishman James Smithson, "for the increase and diffusion of knowledge among men," had by 1876 become the world's leading agency for the international distribution of scientific publications.[2] The history of the formal international publication exchange program, led and managed by the Smithsonian, has been ably detailed by Nancy Gwinn; but less formal mechanisms for distributing scientific publications are not well understood.[3] Due to the prominence of the first secretaries, Joseph Henry and Spencer Baird, the large correspondence networks that became the mainstay of scientific work in the nineteenth century very quickly involved the Smithsonian Institution. Baird, for example, noted in 1860 that he personally wrote about three thousand letters per year, complemented by a large body of additional official correspondence written by staff members on his behalf.[4] This number steadily increased so that by 1875 the office generated upwards of fifty-five hundred letters. Over the course of his more than thirty-five-year employment at the Smithsonian, tens of thousands of letters flowed in and out of his office in the castle headquarters in Washington, D.C., to a vast network of individuals "who looked to him for encouragement, advice, instruction, equipment, supplies, money, and a link to the outside world."[5] This large body of correspondence provides detailed evidence of scientific activities throughout the North American continent and beyond.

Among that correspondence are letters exchanged with Canadians whose scientific endeavors were encouraged by a northward flow of scientific publications to locations as isolated as MacFarlane's Fort Chipewyan in the north and to cities and towns like Toronto; Montreal; Quebec City; and Windsor, Nova Scotia closer to the American border. Information and publications moved both north and south; the flow was decidedly not unidirectional. The relationships among correspondents "were far from being simply exploitative or one-sided," a point Jim Endersby has made about nineteenth-century naturalists in his recent book *Imperial Nature: Joseph Hooker and the Practices of Victorian Science*. Endersby emphasized that correspondence like this reveals "a complex interdependence and mutual benefits, within which individuals bartered whatever they could in an effort to satisfy conflicting desires and competing agendas."[6] In the present essay, archival records at the Smithsonian Institution and elsewhere are probed to demonstrate how the dissemination of knowledge via letters and publications across the Canadian–American border fostered scientific work in both countries.

The pursuit of science in the last half of the nineteenth century had, in the words of Robert Bruce, "become a collective enterprise" where the spread of scientific investigation hinged on the "systematizing of communication" among the growing number of practicing scientists.[7] Several elements figured prominently in this organized communication. Natural history and related scientific societies were formed throughout Canada and the United States by the score. These associations served largely to foster information exchange promoted through lectures given at meetings, journals exchanged widely with other like organizations, and information and publications that accumulated in their museums and libraries.[8] Furthermore, individual scientists obtained scientific information through direct correspondence with fellow practitioners around the world. The correspondence interchange that resulted put scientists and amateurs in direct contact with each other, to the benefit of both metropolis and settler societies in what historical geographer Alan Lester has termed an "imagined geography of empire."[9] Some scholarly attention has focused on particular correspondence networks, or has primarily been concerned with characterizing the correspondents, tracing the flow of particular ideas, or describing the careers of individual scientists, rather than noting the large flow of information and publications that streamed through the networks.[10]

Transmission of information often went beyond elementary correspondence, however, to include the diffusion of many copies of publications. The Smithsonian's first secretary, Joseph Henry, made the importance of communication in science clear in a letter to Thomas Corwin, secretary of the American Treasury on 23 December 1852:

> The value of the rapid diffusion of a knowledge of the discoveries of different countries can scarcely be calculated. It often happens that several persons are engaged in the development of the same truth at the same time and it is therefore of the first importance that each should be informed of the discoveries of the other. . . . Nothing has tended more to retard original research in our own country than the difficulty which has heretofore existed in placing before the scientific world the labors of our own men of science and of obtaining a definitive knowledge of what is doing in foreign countries. It is hoped that our governments will assist in removing obstructions which exist to the free international communication of new truths by a reduction of the high rate of postage on scientific pamphlets, and by the abolition of all duties on printed books.[11]

Views of a similar nature were expressed a few years later on the Canadian side of the border. In his opening address as president of the new Royal Society of Canada, John William Dawson voiced concerns of scientists troubled by the constraints of distance from centers of activity. He lamented that "the public results of Canadian science become so widely scattered as to be accessible with difficulty."[12] Canadian naturalists often faced the "added mortification of finding . . . [their] work overlooked or neglected," he claimed.[13]

The Smithsonian's International Correspondence

To overcome problems of distance or inadequate formalized methods of information exchange, many individuals turned to direct correspondence with authors and organizations. The Smithsonian's correspondence network "was only one of many overlapping networks, but by virtue of its scope and scale . . . [it] brought together an unusually comprehensive and representative sample of the decentralized natural science community."[14] The first two Smithsonian secretaries cultivated an enormous network of correspondents, based mostly in the United States, but extending throughout North America and around the world.[15] The significance of the network from a Canadian perspective can be illustrated by numerous Canadian scientists and individuals in a variety of other walks of life whose letters show up in the archival holdings. The cross-border traffic varied from extensive correspondence between the Smithsonian secretaries and Philip Carpenter and T. Sterry Hunt, both based in Montreal, to a letter or two with naturalists and amateur collectors scattered across the Canadian landscape.[16] Buried within the mass of letters in the Smithsonian Archives are accounts of activities of a large number of individuals captivated by science to some degree.

Baird was one of the greatest natural history collectors of the nineteenth century. Motivated by a desire to build the Smithsonian museum collections, he undertook unrestricted correspondence to support this endeavor. His efforts turned the Smithsonian into warehouses of natural history, mineralogical specimens, and human artefacts. From 1850 to 1878, when he was assistant secretary, the collection grew from about six thousand specimens "to several hundred thousand with an additional 400,000 specimens distributed to other museums and individuals." At his death in 1887, "the holdings totalled over two and a half million specimens."[17] Baird convinced individuals to collect for the Smithsonian by supplying them with collecting tools and reports, books, and scientific journals describing the biological and archaeological items shipped to Washington.

While the Smithsonian's collections increased in size and richness through these efforts, Baird's correspondents derived personal or institutional benefit from their activities. This point is illustrated well in a letter he wrote to J. W. Dawson, principal of McGill University, Montreal, on 6 January 1883, in which he emphasized the value of collection exchanges for McGill's newly opened Redpath Museum:

> I shall be very glad to make the Redpath Museum the depository of the series of Ungava collections we propose to place at the command of the Hudson's Bay Company.
>
> It will be necessary before I can take action formally upon it, to have an official letter from the Company to that effect; as I wish the transaction to go on the records of the Company as well as on those of the Smithsonian Institution.
>
> . . . The National Museum of Washington, and the Redpath Museum of Montreal, ought to be able to render mutual assistance in the way of exchanges of specimens; we having at our command, as you know, vast stores specially illustrating the marine zoology of both the Atlantic and the Pacific coasts. Do you want everything we can send in the way of fishes, invertebrates, &c? Are you prepared to go into the alcoholic business on a large scale, as this, what we

have in greatest abundance? And what can you give us in return, — I do not mean in quantity but in the way of specialities? Good series of well-preserved and accurately identified fossils of Canada, will be welcome, both of plants and animals. Any of the rarer and choicer minerals, especially of large size; the ores of the metals; some of the rarer mammals, and specimens of archaeology and ethnology; all come within our wants.

What can you do in the way of a good series of Canadian buildings stones. . . . We have a magnificent collection of this kind at present from the U.S. and would be glad to extend it to the whole of the northern portions of the continent. . . . I shall be glad to have you make formal applications for any duplicate specimens of marine zoology, &c., if you care to have them; so that they may be duly filed and acted upon at the proper time.[18]

The reciprocity arrangement for specimens, which Baird recommended, is mirrored in the complementary exchange of published and unpublished scientific information that correspondents pursued energetically.[19]

Canadian Correspondents in the Far North

Roderick MacFarland was one such correspondent. By the time he was writing to Baird in the 1870s, Baird had, in cooperation with the Hudson's Bay Company's headquarters, over a decade's experience with its employees collecting for the Smithsonian across northern Canada. From 1859 to 1862, Robert Kennicott, a young American naturalist, had travelled in the Arctic assembling materials for the Smithsonian and recruiting company employees to send specimens south. The company provided Kennicott "with lodging, transportation (of himself and his specimens), and perhaps most importantly, companionship," and Baird corresponded with both Kennicott and collector recruits, shipping supplies north.[20] Thus, MacFarlane's letters were in seasoned hands when they reached the assistant secretary. He relied on Baird not only for Smithsonian publications but also subscriptions to periodicals, which Baird arranged on receipt of funds from MacFarlane.[21] In a letter before Christmas 1875, MacFarlane made his needs clear: "The volume on the Water Birds of your most interesting work on North American birds has not yet reached me. I trust it will soon turn up. My latest Smithsonian Report is for the year 1873. We have also been much, much disappointed at not receiving any of the Harper publications (Weekly, Bazar and Magazine) for the year 1875. . . . Thankly however for the New York Ledger which comes pretty regularly to hand."[22]

Given his far northern location, MacFarlane could not count on speedy transmission of letters to and from Washington; in 1876 letters from Baird took more than seven months to reach MacFarlane.[23] Even so, he continued to inform Baird about publications received (or not) and also send further requests. In November 1876, almost a year after a problem with the Harper subscriptions,

MacFarlane wrote: "Harper publications now reach me pretty regularly thanks to your kind attention to my request. The New York Ledger also comes to hand on the whole well, tho' odd numbers sometimes turn up, some one or more numbers manage to go astray, now and again; but I suppose this is unavoidable all things considered."[24] While satisfied with receipt of periodicals, books and reports had still not arrived at Fort Chipewyan. "I long much to see the Volumes of the Water Birds of your really *great* Work," he wrote. "I *think* a few plates at the end of each volume giving specimens of the eggs would have greatly enhanced the Work to young Naturalists at least." He went on: "I have not seen a Smithsonian Report for an age—we are now very close on *1877*, while the volume for *1873* is the latest on my Book Shelves!"[25] MacFarlane did not hesitate to continue to send additional requests; a few weeks later he asked for several books be purchased and sent north.[26]

While Baird encouraged Hudson's Bay Company traders to supply the Smithsonian with natural history and archaeological collections, many of these men, including MacFarlane, were scientifically inclined in their own right.[27] To support scientific pursuits, they accumulated personal libraries, and the company financed the development of circulating collections at many of the trading posts in the north.[28] Baird and his Smithsonian colleagues had wide-sweeping interests in all North American flora and fauna; so MacFarlane easily gained Baird's assistance in obtaining publications to support his scientific bent. MacFarlane could count on receiving Smithsonian publications as well as on Baird's willingness to handle orders for other books and periodicals.[29] Baird went beyond simply distributing Smithsonian publications to broadly interpret the Smithsonian mandate to increase and diffuse "knowledge among men." As he judiciously encouraged far-flung volunteer collectors to send specimens to the American capital, their information needs were fulfilled with titles from the Smithsonian presses and other sources.

Canadian Correspondents in Applied Sciences

While MacFarlane's scientific interests were secondary to his main line of work, science was more prominent for other Canadians. William Saunders of London, Ontario, for example, was a leading agricultural scientist in the latter decades of the nineteenth century. Born in England in 1836, as a child, Saunders immigrated to Canada with his family. He apprenticed to a druggist in London, Ontario, and soon established his own pharmacy. He early reaped the benefit of scientific societies by joining the American Pharmaceutical Association when he was twenty-four, and by 1877 he was elected president.[30]

His greatest mark in Canadian history arose not from pharmaceutical endeavors but from his agricultural research, which persuaded the Canadian federal government to appoint him as the first director of the newly formed

Experimental Farm System in 1886. His initiatives over the next two decades left an indelible imprint on Canadian agriculture.

Probably stemming from his preoccupation with the pharmaceutical value of plants and his teaching responsibilities at the medical faculty of what would become the University of Western Ontario, Saunders developed an interest in entomology and plant breeding. As early as 1860 he submitted entomological papers to scientific journals, and in 1863 he was one of the founders of the Entomological Society of Canada.[31] When the society launched the *Canadian Entomologist* five years later, Saunders became a major contributor, and he was appointed editor in 1883. Throughout the 1870s he conducted experiments in plant breeding and continued to study insects. The culmination of about two decades' study resulted in a major text, *Insects Injurious to Fruits*, published in Philadelphia in 1883.[32] A first such volume in North America, it was described as "one of the best manuals of the kind" published up to that time.[33] The demand for the volume was sufficient to warrant a second edition in 1892, and Saunders was working on a third at the time of his death.

In contrast to Roderick MacFarlane, Saunders pursued scientific work more or less full time, and for him, more than for MacFarlane, access to scientific publications was very important. He, too, called on the resources of Smithsonian Institution to supply books and reports of interest. When the Entomological Society of Canada was set up in 1863, Saunders was elected secretary. In that capacity, he soon wrote to Secretary Joseph Henry, informing him of the new association and its chief objective, "to form collections of insects for reference and exchange and also an entomological library for the benefit of those interested." Saunders requested "copies of such works on Entomology as you may have on hand." "As our collections will embrace all [taxonomic] orders," he wrote, "we shall particularly need copies of the different lists of North Am[erican] Insects to enable our curator to classify the specimens sent him[.] We shall also be glad to receive any other works you have to spare, now or at any future time bearing on the subject of Entomology even if not *strictly entomological*." He specifically suggested, "Two or three copies of your 'Museum Miscellanea' would be very acceptable and prove a valuable aid in the numbering of the species in our collection." Finally, for his own private library he asked: "If Dr. Le Conte's new catalogue of Coleoptera is out, I should feel grateful for an extra copy for myself. Will part II of Dr. Morris's Synopsis of North Am[erican] Lepidoptera be published this year?"[34]

Saunders's closer proximity to the American capital than MacFarlane's meant a shorter response time could be expected. largely because transportation and the postal system had improved markedly since the middle of the nineteenth century and mail traveled quickly between Canadian and American cities. In fact, within three weeks Saunders acknowledged that the "present of books from the Sm[ithsonian] Inst[itution] for the Entomological Soc[iety] of

Canada . . . [was] at hand with the exception of Leconte's [sic] new species [of]
Col[eoptera]." Still hoping to receive this work, Saunders stated, "We would
be glad if you would forward this to us by post," and he continued to ask for
publications for his own personal purposes: "I should be very glad to receive a
copy of the Synopsis of Diptera for my own use if you have a spare no. left. I
have paid some attention to this family of late and have felt great need of such a
work to assist me."[35]

When Saunders learned about new titles through his network of correspondents, he inquired about obtaining copies from the Smithsonian. A few
months after he had requested a copy of LeConte's work, he wrote to Baird: "I
learn from a recent letter of Mr. Cresson's of Philada [sic] that Dr. Le Conte's
Catalogue of Coleoptera is out also an additional part containing descriptions
of new species. I should feel greatly obliged if you would forward me a copy of
each for myself and one each for our Ent[omological] Society in Toronto." He
repeated a request for "a copy of the Synopsis of Diptera," noting that he
wished a personal copy because "although acting as Secretary for the Society . . .
[he] live[d] 120 miles from its rooms and library so that . . . [he could] seldom
have access to them, and these books are such as one needs constantly for
reference."[36]

Within a few short months, Saunders wrote to Henry in distress to report a
catastrophe: "I regret to inform you that I have suffered much from a disastrous
fire, which occurring in the middle of the night left myself and my family but
barely time to escape with our lives. All of my collections and books were
consumed so that I have to begin afresh again. It is very disheartening to have
the labour of many years consumed in a single night still I feel determined to
commence again and labor hard to regain as speedily as possible my recent
position." By the time he wrote to Henry, Saunders had already called on other
scientists for help in rebuilding his library. Along with sending copies of his
reports and papers, Dr. LeConte encouraged Saunders to write to Henry to
request new copies of publications. "You would confer a great kindness on
me," he stated, "by sending me your Entomological publications again."
Beyond this general plea for new copies he had a very clear view of titles to
replace those he "had so suddenly be[en] deprived of [, namely] Melsheimer's
Cat[alogue] of Coleoptera, Le Conte's Classification, Le Conte's New Catalogue
and 'Description of New Species,' Morris's Cat[alogue] of Lepidoptera and
Synopsis, Hagen's Synopsis of Neuroptera and the Catalogue and Synopsis of
Diptera together with a copy of your Museum Miscellanea."[37]

Over the next few years, Saunders continued to call on the Smithsonian for
publications needed to support his scientific study. On 30 November 1867, for
example, he returned a LeConte title, stating he "was very particular in
explaining that . . . [he] already had a copy of the list which Le Conte had issued
in *March 1863*, and . . . [he] wanted the *later* one embracing all the families

included in his classification of Coleoptera." Throughout the next decade, Saunders bolstered his relationship with the Smithsonian by sending entomological specimens and copies of all the reports and journal of the Entomological Society, and in return he acquired additional Smithsonian publications. By the time Saunders was appointed to head the federal Experimental Farm system in 1886, Joseph Henry had been dead for several years and Spencer Baird would die within the year.

Saunders was near the beginning of his scientific career when he first wrote to the Smithsonian secretary seeking publications to facilitate his studies. Henry and Baird, and other Smithsonian staff members, responded by fulfilling his often very specific requests. His investigations of insects, especially plant pests, were informed by the publications sent from Washington. The value of his *Insects Injurious to Plants*, through two editions, rested in part on ready access to the publications the Smithsonian provided.[38]

In his new position as head of a Canadian research agency, Saunders soon emulated his American correspondents, by setting up an extensive network of recipients of the Canadian Experimental Farm publications. The merits of establishing the experimental farm system (with branches located throughout the country), mirrored by similar research agencies in the United States, were quickly realized. Farmers and others flooded the Central Farm with requests for information. In 1889, for example, only three years after the farm opened, nearly seven thousand letters were received. Saunders and his staff responded by sending out 41,584 copies of reports and bulletins.[39] While he claimed this was a "formidable total of mail matter," a year later the incoming mail had jumped to 17,539 letters, and the staff responded with 19,806 letters and mailed out 218,129 copies of reports and bulletins. Three years on, the corresponding numbers for 1893 were 23,571 letters received, 26,926 letters sent, and 227,899 copies of reports and bulletins mailed out. The steady stream of mailings did not let up. In 1903, the farm received 59,441 letters and mailed out 220,426 reports and 45,485 circular letters. In other words, over the first decade and a half of the operation of the Experimental Farm system, it distributed well over 2.25 million copies of publications across the country and abroad, truly a "formidable total of mail matter."[40] Saunders's experience with the Smithsonian's informal system of distributing publications provided a model for the distribution networks he established as head of a research agency in Canada, although he seems not to have explicitly credited the Smithsonian Institution in this regard.

Canadian Correspondents in the Natural Sciences

Whereas Saunders focussed primarily on agricultural sciences, Henry Youle Hind pursued wider scientific interests. Like Saunders, he was born in England, but a decade earlier.[41] Educated initially in England, Hind also studied in Leipzig for two years between 1837 and 1839, where he may have first gained an interest in chemistry and geology.[42] On his return to England, he studied at Cambridge University for a year, followed by four months in France to improve his French proficiency. In 1846 he moved to North America, settling first in Toronto, where in 1847, at the age of twenty-four, he was appointed to teach science and mathematics at the provincial Normal School. This position launched his career in the sciences, and soon he was publishing on scientific topics, including *Two Lectures on Agricultural Chemistry* in 1850 and *A Comparative View of the Climate of Western Canada* (i.e., Canada West, now Ontario) in 1851.[43] Two years later, he accepted an appointment to the University of Trinity College in Toronto, where he taught in the medical faculty and obtained a master's degree. Soon thereafter, he began a "remarkable outburst of publications on scientific topics and other writings which was to mark the years 1853–1864."[44] His publication initiatives may have been stimulated by his relationship with the Canadian Institute, an association of professional engineers and trained scientists in the Toronto area, and his role as the first editor of the *Canadian Journal: A Repertory of Industry, Science, and Art*, the first scientific periodical in Canada.[45]

The mid-1850s saw an increasing interest in the northwest of British America. Astutely using his connections, Hind was able to secure an appointment as geologist and naturalist to the expedition under Captain John Palliser, which the British government sent to the prairie region of Canada to investigate and report on the material resources, the climate, and the prospects of the construction of a railway. His two reports on his three years of exploration on the western plains were extensive and emphasized his ambition "to explore, to report, to publish."[46]

Over the next few years, Hind attempted to replicate the achievements of his western studies. He undertook explorations in Labrador and British Columbia and published reports on this work. In 1863, he brought out a substantial volume entitled *Eighty Years' Progress of British North America*, which appeared six years later in a new edition, *The Dominion of Canada*.[47] He left Trinity College in 1863 to move to Fredericton, New Brunswick, and then in 1866 he settled in Windsor, Nova Scotia, near the University of King's College, the oldest English-language institution of higher learning in British North America. There he continued to publish; some of his papers appeared in *Nature*, the *Quarterly Journal of the Geological Society of London*, and the *American Journal of Science*.[48] Historian of science Richard Jarrell, concluded that Hind was "a capable geological observer" who "is best remembered as one of the more successful of those British-born, intelligent, active, and restless men who were attracted by Canada's vast empty spaces and great promise."[49] Hind's biographer W. L. Morton, summed up his career: "He was the surveyor, the explorer, the advocate of possibilities. . . . He helped to open the Northwest, itself a considerable achievement for any man."[50]

During his life and work in Nova Scotia in the mid-1870s Hind carried out an extensive correspondence with the Smithsonian secretaries Joseph Henry and Spencer Baird.[51] On 25 September 1873, Hind informed Henry that he was preparing to publish a work on the "Dominion of Canada," and solicited Henry's assistance in procuring "official works relating to the Physical Geography[,] Geology[,] and Climatology of the Northern parts of the United States." Since previously he had acquired copies of the Smithsonian annual reports up to 1861, he requested all subsequent volumes and then itemized further publications of interest:

> Any documents relating to the three subjects I have named would be thankfully received and acknowledged.
>
> I am especially anxious to obtain meteorological observations from the different Hudson Bay posts which I believe you receive, and compress which if such exist I could only get after indefinite delay through the House at London.
>
> I should be glad to get the Land Office Report for 1868–1872 inclusive— and also the Patent Office Reports in the Department of Agriculture for the last five or six years.
>
> Hayden's Geological Report on the Nebraska &c., &c., &c. G.M. Comstock Survey of the Great Lakes—The Mineral Resources of the United States.

To strengthen his request, Hind concluded his letter by reminding Henry that they had met "20 years ago at the Magnetical Observatory at Toronto under the charge of then Lieut. Lefroy, now General Lefroy and Governor of Bermuda."[52]

The Smithsonian correspondence records confirm that Henry acted quickly on receipt of Hind's missive, and his response outlines the type of action that characterized the secretary's desire to promote the diffusion of scientific information:

> Your letter of the 25th Inst. has just been received and I hasten to reply that it will give me much pleasure to supply your desiderata in regard to your proposed work on the Dominion of Canada as far as I am able. The preparation of such a work is worthy of your learning and talents and I doubt not the result will meet your anticipation.
>
> I recollect with much pleasure my visit to Toronto and have ever desired to keep up a friendly relation with the professors of the University at that place. I have received a number of letters of late, all of much interest, from Gen Lefroy and am happy to renew my acquaintance with yourself.
>
> As soon as we can collect the documents mentioned in your letter we shall forward them by express to your address.[53]

Over the next few days Henry wrote to several U.S. government agencies and officials—the commissioner of Agriculture, secretary of the Treasury, and commissioner of the General Land Office (Department of the Interior)—requesting specific publications and others on the three topics Hind had noted: geology, physical geography, and climatology.[54] That fall, Henry sent Hind packages of reports and books, both Smithsonian publications and other government reports as they were received by Henry's office.[55]

Correspondence between Washington and Windsor continued through the 1870s and 1880s. In his role as commissioner of Fisheries for the American government, Spencer Baird had more than a collector's interest in fish species along the Atlantic seaboard.[56] He asked Hind for information about the spawning habits and distribution patterns of coastal species. Throughout 1877, for example, when Hind explored the Labrador shores for the Newfoundland government, steady correspondence between Baird and Hind reveals that scientific information was being passed in both directions.[57] Hind reported on his observations, and Baird sent scientific publications to support his investigations. This information exchange extended over several years. Hind requested and received publications from Baird's office, and in return he sent copies of his own reports and maps to Baird as well as responses to Baird's questions concerning fish stocks and migration patterns in the Atlantic fishery.[58] This reciprocal arrangement proved profitable to both. Baird incorporated information he received from Hind into his fisheries reports, and the publications that arrived at Hind's Windsor home kept him abreast of scientific developments in the United States.

Conclusion

The Smithsonian Institution was a significant node in the correspondence network of Canadian scientists in the latter half of the nineteenth century.[59] Roderick MacFarlane, William Saunders, and Henry Youle Hind were only three of many who corresponded with Joseph Henry and Spencer Baird in Washington. John William Dawson (principal of McGill University, Montreal), Elkanah Billings (palaeontologist with the Geological Survey of Canada, Montreal), L. W. Bailey (University of New Brunswick, Fredericton), Léon Provencher (Quebec naturalist and editor of Le Naturaliste canadien), Robert Scott (a student of native languages in the northwest, located in Red River [now Manitoba]), and J. H. Lefroy and G. T. Kingston (both with the Magnetic Observatory in Toronto) are examples of others who turned to the secretaries of the Smithsonian for scientific publications either issued by the institution itself or made available through the secretaries' efforts.[60]

As the nineteenth century drew to a close, science had reached a new level of activity. Governments financed research agencies, educational institutions adopted new models of teaching and research, and a new professional perspective emerged. These developments on both sides of the Canadian and American border were fostered by the diffusion of scientific information in the form of thousands of copies of reports and books by the Smithsonian. As Daniel

Goldstein has noted in his analysis of American correspondents with Spencer Baird, "the shortage of necessary reference materials was chronic, if not always acute, outside the Eastern cities." The same statement was easily applicable to many Canadian locales of the same period. Although by itself the Smithsonian could not eliminate this problem in either the United States or Canada, it took steps to alleviate the deficiency. Henry and Baird sent out publications that varied from basic collecting instructions to the Smithsonian's own latest monographs published in the *Contributions to Knowledge* and *Miscellaneous Collections* series. The *Annual Report*, highly valued by Canadian correspondents, "included an appendix containing articles on the most important scientific research of the year," sometimes written by Canadians.[61]

Although increasing numbers of commercial publishers and booksellers supplied reading materials to customers in both Canada and the United States as the nineteenth century unfolded, government agencies, of which the Smithsonian Institution is a significant example, offered a parallel and often more important avenue for acquiring scientific publications.[62] Supported by public funds, the agencies turned out scientific pamphlets, reports, maps, and other publications in large numbers and exploited the improving transportation and postal systems to distribute these printed items directly to readers in nearby and remote communities.[63]

Henry and Baird liberally interpreted the mandate of the Smithsonian Institution to "increase" and diffuse "knowledge among men." They willingly corresponded with all comers, whether amateurs or established scientists, recognizing that both could be contributors to the mission of the Smithsonian. The results of their efforts contributed to the maturation of science in the United States and to a lesser extent in Canada. In 1871, Henry claimed that "on account of the wide diffusion of elementary education in the United States, and the greater taste for reading among all classes, there is now greater diffusion of elementary scientific knowledge, and, perhaps more activity of mind directed in the line of scientific thought."[64] Canadians benefited from their proximity to the American capital and the generosity of the Smithsonian secretaries (even if it sometimes took seven months for correspondence to reach more northern locales). The Smithsonian secretaries frequently responded quickly to requests for publications. The northwardly mobile documents, plus the encouragement that would be obtained by a response from an institution of the caliber of the Smithsonian, promoted scientific work in Canada when foundations for this endeavor were being laid down. Whereas the formal exchange program of the Smithsonian fostered the development of a national science in the United States, the informal distribution of publications to Canadians served to promote the careers of individual scientists who could retain their own national identity. Moreover, the reciprocal exchange of information (published and unpublished), biological and mineralogical specimens, and artifacts ensured advancement of science in both Washington, D.C., and numerous communities in Canada. Notwithstanding the difficulties of correspondence reaching isolated northern posts in a timely manner, Canadian scientists of both serious and amateur ranks saw the Smithsonian Institution as a free source of scientific publications. On the American side of the border, the Smithsonian secretaries valued their Canadian correspondents as contributors to expanding collections in the Washington museum.

Notes

I am particularly grateful for the assistance that I received from archivists and librarians at the Smithsonian Institution during a Dibner Library Resident Scholar Fellowship. Further archival work was completed in Canada where I was aided by very helpful archivists and librarians, particularly at Library and Archives Canada, McGill University, Western University, and Dalhousie University. Jennifer Connor, Jonathan Topham, Rima D. Apple, and Suzanne Zeller kindly offered informative insights at various stages of the research. Further suggestions came from an anonymous reviewer. Diane Zerr, a Dalhousie University graduate student, provided additional assistance.

1. R. MacFarlane, Fort Chipewyan, Athabasca, to Spencer Baird, 20 May 1876, Smithsonian Institution Archives, Office of the Secretary (Henry Baird), 1863–79, Incoming Correspondence, RU 52, box 51, vol. 198, no. 74. MacFarlane had written to Baird several months earlier, stating how he had "been much much disappointed at not receiving any of the Harper publications (Weekly, Bazar, and Magazine) for the year 1875. I expect there has been some mistake, and to guard against its recurrence I will now enclose a bill for 2 years subscription." MacFarlane to Baird, 21 December 1875, RU 52, box 51, vol. 198, no. 73. Baird was appointed secretary on 17 May 1878 and served until his death in 1887.

2. On the history of the Smithsonian Institution, especially the significant role of the first two secretaries, see Albert E. Moyer, *Joseph Henry: The Rise of an American Scientist* (Washington, D.C.: Smithsonian Institution Press, 1997), and E. F. Rivinus and E. M. Youssef, *Spencer Baird of the Smithsonian* (Washington, D.C.: Smithsonian Institution Press, 1992).

3. Nancy E. Gwinn, "The Origins and Development of International Publication Exchange in Nineteenth-Century America" (PhD diss., George Washington University, 1996). See also Nancy E. Gwinn, "The Smithsonian Institution Libraries: A Foot in Three Camps," *College & Research Libraries* 50, no. 2 (1989): 206–14, and Nancy E. Gwinn, "The Library of Congress, the Smithsonian Institution, and the Global Exchange of Government Documents, 1834–1889," *Libraries & the Cultural Record* 45, no. 1 (2010): 107–22. Other authors who have discussed international exchange activities include Ilse Sternberg, "The British Museum Library and the Development of the International Exchange of Official Documents," *Electronic British Library Journal* (2002): http://www.bl.uk/eblj/2002articles/article2.html; Maurice C. York, "Alexandre Vattemare's System of International Exchanges in North Carolina," *North Carolina Libraries* 56 (1998): 11–15; Carol Armbruster, "The Origins of International Literary Exchanges: Alexandre

Vattemare's Adventures in America," *Biblion* 5, no. 2 (1997): 128–47; Elizabeth Revai, *Alexandre Vattemare, traite d'union entre deux mondes: Le Québec et les États-Unis à l'aube de leurs relations culturelles avec la France au XIXe siècle* (Montréal: Éditions Bellarmin, 1975); and A. Allardyce, "International Book Exchange," in *Encyclopedia of Library and Information Science*, ed. Allen Kent and Harold Lancour (New York: Marcel Dekker, 1974), 257–77.

4. "Baird maintained contact with his collectors through a voluminous correspondence. During the year 1860, for instance, Baird noted in his journal that he had written, without the aid of a stenographer, 3,050 letters, most of which were undoubtedly to instruct, encourage, and befriend those who could add specimens to the museum." William A. Deiss, "Spencer F. Baird and His Collectors," *Journal of the Society for the Bibliography of Natural History* 9, no. 4 (1980): 639.

5. Daniel Goldstein, "'Yours for Science': The Smithsonian Institution's Correspondents and the Shape of Scientific Community in Nineteenth-Century America," *Isis* 85, no. 4 (1994): 576.

6. Jim Endersby, *Imperial Nature: Joseph Hooker and the Practices of Victorian Science* (Chicago: University of Chicago Press, 2008), 3. Debra Lindsay reached a similar conclusion in her analysis of relationships between the Smithsonian staff and Hudson's Bay Company traders and northern natives; specimens "were exchanged for both economic and extra-economic rewards. Data collecting activities were integrated within existing sociocultural frameworks." Debra J. Lindsay, "Science in the Sub-Arctic: Trades, Trappers and the Smithsonian Institution, 1859–1870" (PhD diss., University of Manitoba, 1989), abstract.

7. Robert V. Bruce, *The Launching of Modern American Science, 1846–1876* (New York: Knopf, 1987), 4.

8. Sarah Gibson, "Scientific Societies and Exchange: A Facet of the History of Scientific Communication," *Journal of Library History* 17, no. 2 (1982): 144–63.

9. Alan Lester, "British Settler Discourse and Circuits of Empire," *History Workshop Journal* 54 (2002): 24.

10. On correspondence networks, see, for example, Raymond Duchesne, "Science et société coloniale: Les naturalistes du Canada français et leurs correspondants scientifiques (1860–1900)," *HSTC Bulletin* 5, no. 2 (1981): 99–139. On characterizing the correspondents, see, for example, Goldstein, "'Yours for Science.'" On tracing the flow of ideas, see, for example, Kennard B. Bork, "Correspondence as a Window on the Development of a Discipline: Brongniart, Cleaveland, Silliman, and the Maturation of Mineralogy in the First Decades of the Nineteenth Century," *Earth Sciences History* 18, no. 2 (1999): 198–245; Ian Campbell and David Hutchison, "J. D. Forbes' Scientific Correspondence with Switzerland," *Scotia* 8 (1984): 26–42; and also Susan Sheets-Pyenson, "Geological Communication in the Nineteenth Century: The Ellen S. Woodward Autograph Collection at McGill University," *Bulletin of the British Museum of Natural History*, Historical series, 10, no. 6 (1982): 179–226. On the careers of particular scientists, see, for example, D. F. Branagan, "Alfred Selwyn—19th Century Trans-Atlantic Connections via Australia," *Earth Science History* 9, no. 2 (1990): 143–57.

11. Joseph Henry to Thomas Corwin, 24 December 1852, in *The Papers of Joseph Henry*, vol. 8, *January 1850–December 1853: The Smithsonian Years*, ed. Marc Rothenberg et al. (Washington, D.C.: Smithsonian Institution Press, 1998), 418.

12. For another American example, see letters written by Leo Lesquereux, an American paleobotanist located in Columbus, Ohio, to J. W. Dawson, 14, 31 March 1861, John William Dawson Papers, McGill University.

13. John William Dawson. "[Opening Remarks, Inaugural Session 1882]," *Proceedings and Transactions of the Royal Society of Canada for the Years 1882 and 1883* 1 (1883): viii–ix.

14. Goldstein, "'Yours for Science,'" 576.

15. The Joseph Henry Papers project has published eleven large volumes of Henry's correspondence, which contain about ten percent of his letters (conversation with the editor, Marc Rothenberg, July 2001).

16. Canadian scientific correspondents included L. W. Bailey (Fredericton, New Brunswick), Robert Bell (Montreal), Charles Bethune (Port Hope, Ontario), Elkanah Billings (Montreal), Philip Carpenter (Montreal), William Cooper (Ottawa), Edward Cortland (Ottawa), N. F. Dupuis (Montreal), William Hincks (Toronto), Henry Youle Hind (Windsor, Nova Scotia), David Honeyman (Halifax), Henry How (Windsor, Nova Scotia), T. Sterry Hunt (Montreal and Boston), William Brydon Jack (Fredericton), G. T. Kingston (Toronto), R. W. M. Lachlan (Montreal), George Lawson (Kinston, Ontario), J. H. Lefroy (Toronto), William Logan (Montreal), G. F. Matthew (Fredericton), Léon Provencher (Cap Rouge, Quebec), William Saunders (London, Ontario), A. R. C. Selwyn (Montreal), Charles Smallwood (Montreal), Henry Vennor (Montreal), J. F. Whiteaves (Montreal and Ottawa), and Daniel Wilson (Toronto).

17. Deiss, "Spencer F. Baird and His Collectors," 635.

18. Spencer Baird to J. W. Dawson, 6 January 1883, RU 33, vol. 135, nos. 203–7.

19. On scholarly study of gift exchange and reciprocity see Aafke E. Komter, ed., *The Gift: An Interdisciplinary Perspective* (Amsterdam: Amsterdam University Press, 1996), and on its ancient roots of gift exchange, see Marcel Mauss, *The Gift: The Form and Reason for Exchange in Archaic Societies*, trans. W. D. Halls (New York: Norton, 1990).

20. Deiss, "Spencer F. Baird and His Collectors," 642. Deiss writes: "Baird was responsible for seeing to it that they received all necessary collecting equipment and supplies as well as gifts designed to keep them happy. Even though Baird had to send alcohol for preserving specimens in metal cans or tin-lined kegs to prevent tapping by thirsty company employees, he made sure that they received plenty of good whiskey and reading matter, such as *The way to do good, Fox's book of Martyrs,* and *Walker on women.*"

21. It seems that MacFarlane sent a sum of money to Baird, which could be used for subscriptions, for he wrote in July 1875, "but as the money from Mr. Gardiner and myself has been paid over to you, you would much oblige us by investing it in such a manner as you may deem most advantageous." MacFarlane to Baird, 27 July 1875, RU 52 box 51, vol. 198, no. 72.

22. MacFarlane to Baird, 21 December 1875, RU 52, box 51, vol. 198, no. 73.

23. MacFarlane to Baird, 24 May 1876, RU 52, box 51, vol. 198, no. 74.

24. MacFarlane to Baird, 25 November 1876, RU 52, box 51, vol. 198, no. 75.

25. MacFarlane to Baird, 25 November 1875, RU 52, box 51, vol. 198, no. 75.

26. MacFarlane to Baird, 21 December 1876, RU 52, box 51, vol. 198, no. 76.

27. See Greg Thomas, "The Smithsonian and the Hudson's Bay Company," *Prairie Forum* 10, no. 2 (1985): 283–305.

28. See, for example, Michael R. Angel, "Clio in the Wilderness; or, Everyday Reading Habits of the Honourable Company of Merchant Adventurers Trading into

Hudson's Bay," *Manitoba Library Association Bulletin* 10, no. 3 (June 1980): 14–18; Judith Hudson Beattie, "'My Best Friend': Evidence of the Fur Trade Libraries Located in the Hudson's Bay Company Archives," *Épilogue* 8, nos. 1 & 2 (1993): 1–32; Fiona A. Black, "Beyond Boundaries: Books in the Canadian Northwest," in *Across Boundaries: The Book in Culture & Commerce*, ed. Bill Bell, Philip Bennett, and Jonquil Bevan (Winchester, England: St. Paul's Bibliographies, 2000), 91–115; Leslie D. Castling, "Peter Fidler's Books," *Manitoba Library Association Bulletin* 11, no. 4 (1981): 47–48; Jean Murray Cole, "Keeping the Mind Alive: Literary Leanings in the Fur Trade," *Journal of Canadian Studies* 16, no. 2 (Summer 1981): 87–93; Debra Lindsay, "Peter Fidler's Library: Philosophy and Science in Rupert's Land," in *Readings in Canadian Library History*, ed. Peter F. McNally (Ottawa: Canadian Library Association, 1986), 209–29; Michael Payne and Gregory Thomas, "Literacy, Literature, and Libraries in the Fur Trade," *Beaver Outfit* 313, no. 4 (Spring 1983): 44–53; and Laura J. Murray, "The Uses of Literacy in the Northwest," in *History of the Book in Canada*, vol. 1, *Beginnings to 1840*, ed. Patricia Lockhart Fleming, Gilles Gallichan, and Yvan Lamonde (Toronto: University of Toronto Press, 2004), 187–93.

29. Some of MacFarlane's books have survived. See Beattie, "'My Best Friend,'" 16–32.

30. The most detailed biography of Saunders is Elsie M. Pomeroy, *William Saunders and His Five Sons: The Story of the Marquis Wheat Family* (Toronto: Ryerson Press, 1956), esp. 4–5. See also Ian M. Stewart, "Saunders, William," *Dictionary of Canadian Biography* [online]. http://www.biographi.ca/009004-119.01-e.php?&id_nbr'7688&&PHPSESSID=ni13m6acvcdi15acvinlud2.

31. J. T. H. Connor, "Of Butterfly Nets and Beetle Bottles: The Entomological Society of Canada, 1863–1960," *HSTC Bulletin* 6, no. 3 (1982): 151–71.

32. William Saunders, *Insects Injurious to Fruits* (Philadelphia: J. B. Lippincott & Co., 1883).

33. Charles J. S. Bethune, "The Rise and Progress of Entomology in Canada," *Proceedings and Transactions of the Royal Society of Canada*, 2nd ser. 4 (1898), section 4, 163. Saunders's book "remained the primary reference in that field for many years." T. H. Anstey, "The Formation of the Experimental Farms," *Prairie Forum* 11 (Fall 1986): 190.

34. William Saunders to Joseph Henry, 1 May 1863, RU 26, box 12, vol. 55, no. 231.

35. William Saunders to Joseph Henry, 21 May 1863, RU 26, box 12, vol. 55, no. 280.

36. William Saunders to Spencer Baird, 8 November 1863, RU 26, box 12, vol. 55, no. 274. Saunders had been informed by Cressen that he should include funds to cover the postage of the books to Canada. He included $2.00 for this purpose, but a note on the letter indicates the money was returned.

37. William Saunders to Joseph Henry, 14 April 1864, RU 26, box 12, vol. 52, no. 319.

38. The preface to *Insects Injurious to Plants* acknowledges works by numerous authors that Saunders had consulted, copies of which he had obtained from the Smithsonian Institution.

39. William Saunders's report in *Experimental Farms Reports . . . for 1889* (Ottawa: Printed by Brown Chamberlin, Queen's Printer, 1890), appendix to the Report of the Minister of Agriculture, Sessional Papers, no. 63, 1890, 5.

40. The numbers of incoming and outgoing letters and publications distributed by the Central Experimental Farm are given in the director's annual reports. Further discussion about this correspondence is given in Bertrum H. MacDonald, "Getting the Books Published: The Instrumental Role of Governments and Their Agencies," *Facsimile*, no. 19 (1998): 13–18.

41. The most comprehensive biography of Hind is W. L. Morton, *Henry Youle Hind, 1823–1908*, Canadian Biographical Studies series, no. 7 (Toronto: University of Toronto Press, 1980). See also Richard A. Jarrell, "Hind, Henry Youle," *Dictionary of Canadian Biography* [online] http://www.biographi.ca/009004-119.01-e.php?&id_nbr=6783&interval=25&&PHPSESSID=fqfsvo7rdbapc47naavooomm5.

42. Morton, *Henry Youle Hind*.

43. Henry Youle Hind, *Two Lectures on Agricultural Chemistry* (Toronto: H. Scobie, 1850); and *A Comparative View of the Climate of Western Canada Considered in Relation to Its Influence upon Agriculture* (Toronto: Brewer, McPhail, 1851).

44. Morton, *Henry Youle Hind*, 21.

45. Jarrell, "Hind, Henry Youle."

46. Morton, *Henry Youle Hind*, 30–33, 83.

47. Henry Youle Hind, *Eighty Years' Progress of British North America Showing the Developments of Its Natural Resources, by the Unbounded Energy and Enterprise of Its Inhabitants . . . with a Large Amount of Statistical Information from the Best and Latest Authorities* (London: S. Low & Marston; Toronto: L. Stebbins, 1863), and *The Dominion of Canada; Containing a Historical Sketch of the Preliminaries and Organization of Confederation . . . with a Large Amount of Statistical Information, from the Best and Latest Authorities* (Toronto: L. Stebbins, 1869).

48. Henry Y. Hind, "The Figure of the Earth in Relation to Geological Inquiry," *Nature* 10 (2 July 1874): 165–67; Henry Y. Hind, "Observations on Supposed Glacial Drift in Labrador Peninsula, Western Canada, and on the South Branch of the Saskatchewan," *Quarterly Journal of the Geological Society of London* 20 (February 1864): 122–30; Henry Y. Hind, "On Two Gneissoid Series in Nova Scotia and New Brunswick, Supposed to be Equivalent of the Huronian (Canadian) and Laurentian," *Quarterly Journal of the Geological Society of London* 26 (January 1870): 468–78; and Henry Youle Hind, "On the Laurentian and Huronian Series in Nova Scotia and New Brunswick," *American Journal of Science and Arts*, 2nd ser. 49, no. 147 (May 1874): 347–55.

49. Jarrell, "Hind, Henry Youle."

50. Morton, *Henry Youle Hind*, 123.

51. Neither Jarrell nor Morton seems to have been aware of this correspondence when they prepared their biographies.

52. Henry Youle Hind to Joseph Henry, 25 September 1873, RU 26, box 54, vol. 137, no. 214. Henry held Lefroy in high regard. At the time he was in Toronto, Lefroy was one of the leading physical scientists on the continent, when the discipline was not well established in the United States. See Gregory A. Good, "Toronto Magnetic Observatory and International Science ca. 1850," *Vistas in Astronomy* 28, part 1 (1985): 387–90, and Gregory A. Good, "Between Two Empires: The Toronto Magnetic Observatory and American Science before Confederation," *Scientia Canadensis* 10, no. 1 (1986): 34–52.

53. Joseph Henry to Henry Y. Hind, 30 September 1873, RU 33, vol. 35, nos. 714–15.

54. Joseph Henry to F. Watts, Commissioner of Agriculture, 3 October 1873, RU 33, vol. 35, nos. 743–44; Joseph Henry to W. A. Richardson, Secretary of the Treasury, 3 October 1873, RU 33, vol. 35, nos. 750–51; and Joseph Henry to Milles Drummond, Commissioner of the Lands Office, 3 October 1873, RU 33, vol. 35, nos. 752–53.

55. Henry Y. Hind to Joseph Henry, 15 November 1873, RU 26, box 54, vol. 137, no. 18; Joseph Henry to Henry Y. Hind, 5 December 1873, RU 33, vol. 36, nos. 637–38; and Joseph Henry to Henry Y. Hind, 15 December 1873, RU 33, vol. 36, no. 782.

56. See Dean C. Allard, "Spencer Baird and the Scientific Investigation of the Northwest Atlantic, 1871–1887," *Northern Mariner* 7, no. 2 (1997): 31–39, and Dean C. Allard, "Spencer Baird and Support for American Marine Science, 1871–1887," *Earth Sciences History* 19, no. 1 (2000): 44–57.

57. In 1877, in particular, Hind wrote to Baird frequently, sometimes twice per month.

58. The reciprocal exchange of information and publications between Hind and Baird is confirmed by numerous letters throughout the period from 1877 through 1882 in the Smithsonian Institution Archives.

59. Growing correspondence activities were aided by the publication of directories after the mid-1860s. See Vernon N. Kisling Jr. "*The Naturalists' Directory* and the Evolution of Communication among American Naturalists," *Archives of Natural History* 21, no. 3 (1994): 393–406.

60. Correspondence between John William Dawson and the Smithsonian secretaries occurred through the 1860s, 1870s, and 1880s. See Joseph Henry to Elkanah Billings, 31 December 1870, RU 33, vol. 22, nos. 383–84; Joseph Henry to Elkanah Billings, 23 January 1871, RU 33, vol. 22, nos. 676–77; Joseph Henry to Elkanah Billings, 31 March 1891, RU 33, vol. 24, no. 489; Elkanah Billings to Joseph Henry, 2 March 1871, RU 26, box 36, vol. 108, no. 153; and Elhanah Billings to Joseph Henry, 23 May 1871, RU 26, box 36, vol. 108, nos. 154–55. Correspondence between Bailey and the two Smithsonian secretaries spanned 1862 to 1880. Provancher corresponded with Henry in the late 1860s. See Robert Scott to Joseph Henry, 30 December 1872, RU 26, box 49, vol. 131, no. 164; Robert Scott to Joseph Henry, 15 January 1873, RU 26, box 49, vol. 131, no. 165; and Robert Scott to Joseph Henry, 2 June 1873, RU 26, box 49, vol. 131, no. 166. Numerous letters between Henry and Lefroy and Henry and Kingston, spanning the 1850s through the 1870s, are found in the Smithsonian Institution Archives.

61. Goldstein, "'Yours for Science,'" 588.

62. David Hall discusses specialized scientific publishing and the Smithsonian Institution in "Erudition and Learned Culture," in *A History of the Book in America*, vol. 2, *The Industrial Book, 1840–1880*, ed. Scott E. Casper et al. (Chapel Hill: Published in Association with the American Antiquarian Society by the University of North Carolina Press, 2007), 347–60. On bookselling in Canada, see Greta Golick, "Bookselling in Town and Country," in *History of the Book in Canada*, vol. 2, *1840–1918*, ed. Yvan Lamonde, Patricia Lochhart Fleming, and Fiona A. Black (Toronto: University of Toronto Press, 2005), 210–26.

63. The significant role governments played in publishing, including scientific publications, is recounted in Scott E. Caster, "The Census, the Post Office, and Government Publishing," in Casper et al., *History of the Book in America*, 2:178–93; Charles A. Seavey with Caroline F. Sloat, "The Government as Publisher," in *A History of the Book in*

America, vol. 4, *Print in Motion: The Expansion of Publishing and Reading in the United States*, ed. Carl F. Kaestle and Janice A. Radway (Chapel Hill: Published in Association with the American Antiquarian Society by University of North Carolina Press, 2009), 260–75; and Marcel Chotkowski LaFollette, "Crafting a Communications Infrastructure: Scientific and Technical Publishing in the United States," in Kaestle and Radway, *A History of the Book in America*, 4:234–59. Further discussion about the availability of scientific publications in Canada is given in Bertrum H. MacDonald, "Science and Technology," in *History of the Book in Canada*, vol. 2, *1840–1918*, ed. Yvan Lamonde, Patricia Lochhart Fleming, and Fiona A. Black (Toronto: University of Toronto Press, 2005), 423–28.

64. Goldstein, "'Yours for Science,'" 583, quoting the *Annual Report* for 1871.

Writing Medicine

George M. Gould and Medical Print Culture in Progressive America

JENNIFER J. CONNOR

Just when the twentieth century was dawning, in December 1899, an American medical editor told his audience about "the shocking abuses that have sprung up in the realm of medical journalism." By recounting "evil tendencies" of journals owned by proprietary medicine manufacturers and self-promoting medical men, he maintained, "it will thus be seen that medical journalism is in a state of chaos."[1] A century later, this message of concern has changed little — and may still reverberate through the language of abuse and harm: "The whole business of medical journals is corrupt because owners are making money from restricting access to important research, most of it funded by public money," pronounced former *British Medical Journal* editor Richard Smith in his book, *The Trouble with Medical Journals*.[2] A self-styled iconoclast, Smith dissected the problematic state of late-twentieth-century medical journals in their peer review; authorship; editorship; research; and relationships with patients, media, and sponsors. His highly personal account echoes essays published in 2000, *Ethical Issues in Biomedical Publication*, edited by Anne Hudson Jones and Faith McLellan; this collection demonstrates that a powerful voice of alarm belongs to editors of medical journals: four of the fourteen essayists are or were editors of the most influential journals in medicine, and two were assistant editors.[3] Indeed, the "Uniform Requirements for Manuscripts Submitted to Biomedical Journals," developed by the International Committee of Medical Journal Editors in 1978 and adopted by several hundred medical journals, continues to evolve as editors around the world demand that someone take public responsibility for the content and form of a written medical communication.[4]

Despite this vocal and sustained concern about medical journals over the course of the twentieth century, journals have not received extensive attention from historians.[5] Little is therefore known about them within the context of print culture—whether the topic be the role of editors in their production and their reform; the professionalization of the medical editor; the nature of medical authorship; or the impact of external forces on the publication process. Still less is known about other forms of medical publication, such as the book, from the perspective of book history and print culture: instead of synthetic studies, such subjects as medical publishers or author-publisher relations have been relegated—as have their journal history counterparts—mainly to individual examples in chronological, bibliographical, or biographical accounts.[6]

Yet, as with studies by those actively involved in medical journals today, these topics were evident in the 1899 paper read by editor P. Maxwell Foshay before the Northwestern Ohio Medical Association and subsequently published in the *Journal of the American Medical Association*. As he outlined ways to distinguish good journals from bad, his emphasis that "preference should always be given to journals that are owned and entirely controlled by medical men" implies experience with other kinds of journal ownership. Equally significant for Foshay was the editor's role: "Know something about the character and reputation of the editor of the journal for which you subscribe," he advised. "It is not in reason to expect a clean journal from the hands of a self-advertiser or of a man of doubtful reputation, nor on the other hand is it likely that a physician of good character will edit an immoral journal." Foshay, who edited the relatively minor *Cleveland Journal of Medicine*, seemed here to recite the mantra of George M. Gould, an American medical editor of international stature who had been actively and widely promoting causes in aid of medical literature and its control by the medical profession. So well known was he at the time that Foshay's listeners may have recognized and understood that it was Gould who exemplified the reputable, trustworthy medical editor; indeed, the vocabulary of "abuses," "evil," "chaos," "clean," and "immoral" itself, as we shall see, suggests that Gould and his many communications had deeply influenced Foshay.

Although George Gould had been a national medical editor for a few years before this time, he had recently increased his profile and stepped up his campaign for improved medical literature and its collection. His condemnations of medical publishers who refused to permit wider circulation of their books and journals through gifts to libraries and reprinting in textbooks were printed as circulars, published in medical journals, presented at medical associations around the North American continent, and subsequently drew published comment from editors of elite journals everywhere, such as the British *Lancet*— all in addition to his own editorials in his journals. In 1898 Gould had established a new society, the Association of Medical Librarians, to conduct an exchange of medical literature to improve the state of medical libraries—and this society gave him an additional platform from which to criticize publishers and proselytize for medical journals both owned and controlled by the medical profession. He immediately addressed the American Medical Editors' Association meeting, exhorting its members to perform their proactive duty.[7] It is therefore not inconceivable that Dr. Foshay may have been in attendance and would have been convinced to spread the word among his own constituency in Ohio; if not there, then he might have become aware of Gould and his campaign at the American Medical Association (AMA) conference itself, when the AMA adopted Gould's resolution to encourage the organization and support of public medical libraries in all American communities and to give its own journal free to library members of his new Association of Medical Librarians.[8] Certainly, Foshay and colleagues would know about Gould's campaign, which received notices in the *Journal of the American Medical Association* as well.

As a central figure in raising awareness of many issues in medical publishing at the end of the nineteenth century, George M. Gould thus serves historically as a powerful lens through which to view the medical profession's concern about ownership and distribution of medical knowledge. One of the most outspoken of leading American medical editors in the Progressive Era, from 1898 to 1906 Gould edited three weekly medical journals. He editorialized on every imaginable topic of his time, had broad publishing activities that embraced most facets of print culture, and was a driving force in many of them. For him, the process of "writing" medicine above all meant "righting" medicine. In this regard, his routine work with prominent medical publishers and his correspondence with medical leaders reveal contests between publisher and medical profession, contests usually portrayed as ethical. After brief biographical information to provide context for Gould's lifelong involvement in and commitment to print culture, this essay therefore describes his medical publishing activities according to his views on the production and distribution of medical literature. As it suggests, Gould was a consummate medical editor, of international stature, who articulated concerns virtually identical to those of editors around the globe today; however, where his present-day counterparts could adopt an orchestrated approach to issues in medical publishing through their international committee, according to his own contemporaries Gould acted either as sentinel or as one who led the charge into battle for the profession.

George M. Gould: A Life in Print

A successful Philadelphia-based ophthalmologist, George Milbry Gould (1848–1922) once admitted that his first choice of vocation was that of librarian.[9] As an activist for books and libraries instead, he promoted free distribution of

Figure 11. George Milbry Gould, A.M., M.D., 1848-1922. From Irving A. Watson, ed., *Physicians and Surgeons of America: A Collection of Biographical Sketches of the Regular Medical Profession* (Concord, NH: Republican Press Association, 1896).

medical and scientific literature, and he established an Association of Medical Librarians to this end (still active as the Medical Library Association). His added experience as bookseller, stationery salesman, researcher, inventor, and publisher meant that he developed a unique, in-depth understanding of print culture in the late nineteenth century.

In addition to memorial statements, reviews of his books describe Gould's skill as a writer: he "is not always right, but he is always interesting"; "a man of strong opinions who expresses them with a vigorous and unusual frankness"; he needed no introduction to the "reading medical profession as an erudite, trenchant writer," for "no one can forget his terse and enlightening editorials in American Medicine and other medical journals which he has edited."[10] A review of his book of poetry, published in 1897, crystallized features of his persona as writer: "In his prose writings we seldom see his sympathetic side. He writes prose to correct some abuse."[11] Indeed, Gould's contemporaries agreed that his "white-hot zeal" and "recklessness of the pioneer" drove him both to write and to publish feverishly "to correct some abuse." The publication output of this "dynamically overcharged" man (as one friend saw him) therefore is daunting.[12] He was editor of three national weekly medical journals, *Medical News* (1891–95), *Philadelphia Medical Journal* (1898–1901), *American Medicine* (1901–6); an encyclopedia of medicine; and an annual volume on medicine and surgery. He was author and coauthor of dozens of pamphlets; hundreds of editorials and notes; a compilation of selections on topics from vivisection to immortality; a book on eye diseases; a primer for medical writers; a collection of poetry; biographies of writer Lafcadio Hearn and poet Edmund Stedman; several volumes of clinical biographies of famous men; a book on the meaning of life; a two-volume history of Jefferson Medical College; a book titled *Anomalies and Curiosities in Medicine*; and medical dictionaries that would remain part of the medical literature for much of the twentieth century, eventually retaining his name as *Blakiston's Gould Medical Dictionary*. His publishers include all the Philadelphia firms that specialized in medicine: Lea, Saunders, Blakiston, and Lippincott.

Gould's devotion to books is evident throughout his unusual life. He decided at the age of ten to study for the ministry; after the Civil War, he attended Ohio Wesleyan University (eventually receiving A.B. and A.M. degrees), then entered Harvard Divinity School in 1869, interrupting his studies to take in lectures in history, theology, and literature at the universities of Berlin and Leipzig from 1871 to 1873. He graduated from Harvard after returning for another year. Soon after, he confided that "owing to weak throat and lungs, poor eyes and a rebellious intellect," he would not succeed as a minister. After a stint as book-keeper in the family coal and salt business and a failed application to be librarian at the state library in Columbus, Ohio, he moved to Chillicothe to co-run two bookstores. Along the way, he learned telegraphy and attended literary clubs to relieve his boredom. Before his first partnership ended abruptly, his store had a successful line of sales in chromolithographs. With another partner, he operated a second store and may have published a book under their name—Gould and Kello. Within a couple of years, this store was destroyed by fire, leaving him to work briefly in a railway car foundry, before this business too failed. For the next three years, Gould worked as a traveling salesman for a New York–based firm of blank books and diaries, spending spare time reading in the Ohio State library; he toyed with the idea of writing fiction, and increasingly turned his attention to medicine out of frustration with physicians.

Gould studied medicine at Jefferson Medical College in Philadelphia, and upon graduation in 1888, at age forty, he received honorable mention for his thesis.[13] By then, he had published pamphlets, coauthored the first of his books, and begun his dictionary work. *A New Medical Dictionary* appeared in 1890, with title changes published to at least 1906. *A Pocket Medical Dictionary* appeared in 1892, followed by *An Illustrated Dictionary of Medicine, Biology, and Allied Sciences* in 1894 with a supplement in 1904. Under the title *Gould's Medical Dictionary*, others continued the work. All these dictionaries were favorably reviewed in the medical press and placed on required lists for medical students. By Gould's death in 1922, half a million copies had been sold; the second edition of *Gould's Medical Dictionary*, published by Blakiston's in 1928, stated that more than seven hundred thousand copies had been distributed around the world. (The last edition appears to be the fourth, published by McGraw-Hill in 1979 as *Blakiston's Gould Medical Dictionary*.) However, probably Gould's most lasting contribution is the often-reprinted—and as any quick Internet search will reveal, still widely circulated—*Anomalies and Curiosities of Medicine*.

With this prolific output, George M. Gould serves well as a historical lens through which to view medical communication in Progressive America. Although his work was ceaseless and reflected a seamless worldview to him, to understand its significance, it is useful to separate his activities and consider them individually. Inevitably, Gould wrote repeatedly and extensively on each activity, often for different audiences, and a thematic perspective allows us to see that his work centered on four main aspects of production and distribution of medical literature: centralizing its exchange; controlling its publication; expanding its collection; and standardizing its language and form. Fundamental to all are notions of progress in and ownership by the medical profession.

Centralizing the Exchange of Medical Literature

Libraries routinely exchanged material among themselves, but to bring order to the chaos of continually proliferating medical literature George Gould and medical colleagues around the United States took up the project in a major way.[14] In 1894 Gould declared that time, money, and medical literature were wasted when physicians purchased their own libraries, only to see them destroyed or dispersed at the end of life and career. He called for common libraries to be established for the profession and for expansion of libraries already founded; he believed a centralized exchange of medical literature among libraries was the primary way to achieve these goals.

In 1896, between editorial engagements, Gould found time to establish such an exchange. This venture relied on the cooperation of the *Journal of the American Medical Association*, which published the lists of duplicates he received from librarians and physicians. With only small expense on his part for postage and typing, Gould announced a year later not merely that he had helped several libraries fill their shelves but also that he had "rescued, preserved, and for all time, placed at professional disposal" upward of six thousand items.[15] His one-man operation had thus achieved such "unexpected" success that he had to consider alternative arrangements for its continuation. Believing in united professional effort and international outreach, he suggested that a corporate organization, preferably a government agency, should take over the endeavor: in his view, the Surgeon-General's Office was ideally suited to act as clearing house for medical literature in much the same way as the Smithsonian Institution served the scientific community.

By 1898, Gould's vision of a centralized clearing house shifted to a voluntary organization such as a medical society. Initially he worked in tandem with another colleague, C. D. Spivak, who arranged for a new AMA committee on libraries. At the AMA conference in June, Gould chaired the first meeting of this committee, with Spivak as secretary. This meeting attracted twenty-five physicians and adopted Gould's proposed resolution; then passed at the AMA's general session, the resolution sanctioned any legitimate method of promoting medical libraries while urging AMA members to aid in the cause. At heart, however, Spivak and Gould diverged in their goals to improve medical libraries: where Spivak preferred a community-based scheme within the AMA, Gould envisaged an organization that could act on an international level. After publishing a small periodical for three years, *Medical Libraries*, Spivak eventually left the field, owing in large measure to Gould's superior performance.[16]

Where Spivak had an idea, Gould had a vision. It did not include the AMA or the American Library Association, mainly because Gould sought independent medical control over the operation. Indeed, while supporting the AMA meeting he had simultaneously acted on a suggestion that the profession needed a new society: only one dedicated to medical libraries would alleviate pressure on them to stay current. This idea, he acknowledged, seemed to have "sprung up in the minds of several at once."[17] Because he was once again too busy to undertake the work alone, having recently founded the *Philadelphia Medical Journal*, he had issued invitations to meet in May 1898 to set in motion such an association.

In his address at this separate meeting in Philadelphia, Gould outlined tasks to expand and improve the quality of "public" (as opposed to personal) medical libraries. Above all, he sought a systematic exchange of library duplicates so that "volumes not needed by one library may by exchange or purchase find their way to the library heretofore without them."[18] One year later, the association had raised nearly half its goal of $1,000 to run the exchange (the largest donation being $100 from Gould himself), and within another few months it had enough funds to proceed. The Exchange of the Association of Medical Librarians began a short time later in Philadelphia and continues today through the Medical Library Association.[19]

Controlling the Publication of
Medical Literature

Establishing a mechanism for sharing medical literature, along with a professional society, that both remain active and strong over a century later (the Medical Library Association has more than four thousand members) are perhaps Gould's greatest legacy. At the time, however, he was better known for his role as editor. After editing *Medical News* for five years, he received many letters of regret from colleagues, ordinary and eminent alike, at his departure in 1895. "I am but a young, obscure Country Physician," wrote one, "but in your retirement from the 'Medical News' I feel a keen loss. Your Editorials do not only reflect your noble soul and profound medical knowledge, but they are literary gems to me. Collected and compiled in [one] volume they would make a priceless book."[20] Explained the renowned Alfred Stillé, "As an Editor you are stationed on one of the watch-towers of the medical citadel, and I think it must please all loyal physicians to note that you have not failed to warn the garrison of the enemies without as well as of the traitors (sometimes unconsciously so) within the walls."[21] Again and again, letters and editorials emphasized his "fearless determination to do and to uphold the right," his stand on all questions "upholding the honor and dignity of the profession," and his turning the *News* into a "power for good," the "ideal of a perfect up to date medical journal."[22] His friends later recalled how he handled the "staid, reliable, Quaker-like old" journal in stellar fashion.[23]

Gould's primary aim as editor was to improve medical education, medical science, and the medical profession, but some of his sustained campaign was to represent colleagues who believed the medical communications circuit— especially publication, distribution, and access—should be controlled by the medical profession. A letter from Dr. William Osler, his friend, fellow supporter of medical journals and libraries, and the most well-known physician of the time, delineated this concern for control:

This is a bit of news, indeed, about the News. I am very sorry. You have done splendid work for the profession, work which should be continued. In many ways too it is rather a serious loss for Phila. I do not like to see the old town without a good "weekly." We need a good independent journal. I would willingly subscribe $1000 during 1896 to help organize one if you had the editorial control. Saunders I suppose would like to have one but there is the same objection to him as to Messrs Appleton [Osler's own publisher in New York], Wood, Lea & Co.[24]

In other words, the publisher then had ultimate control.

Nevertheless, Gould does not seem to have suffered much editorial interference from the publisher of *Medical News*. Indeed, one letter from Charles M. Lea indicates that the publisher could only suggest changes to editorials—in as tactful and hopeful a manner as possible. Questioning one on the "G.T.R. Doctors," Lea felt that Gould's term "business doctors" (presumably because they were company doctors for the Grand Trunk Railroad) would create "ill feeling" among prominent medical men—"men to whom both you & I must look for literary aid & support"—and Lea merely asked him to reconsider.[25] "Anyhow, my dear doctor, will you not postpone the appearance of the one in question, take time to quietly consider its present & ultimate effect upon yourself & the interests placed in your hands & let me have the pleasure of hearing from you," he gently proposed. It is true that a year later Lea did intervene by asking the journal's printer not to proceed with preparation of an editorial for publication. However, the context was somewhat different on this occasion. As he politely explained, this decision was based on a previously understood agreement with Gould that the journal not enter into local politics. Lea had submitted Gould's editorial to his partners, who agreed with the policy that, "as a publishing house, we stand aloof in all such controversies."[26] The matter in question related to "internal troubles" of Jefferson (its governance), not to a medical issue, and the fact that Gould had addressed a separate thirty-page typewritten diatribe to the trustees of Jefferson about their mismanagement probably reveals personal motives at play in his proposed editorial for *Medical News*.

Gould remained editor of *Medical News* for another two years, before Lea moved it to New York (apparently for business reasons). He would fare less well with a publisher in his next venture. In 1896, he corresponded with W. B. Saunders, the Philadelphia-based publisher of his *Year-Book*, about supporting a journal. Saunders regretfully turned him down, stating that since such an enterprise required a guarantee of $25,000, Gould might raise the money through a stock company; Saunders would then assume publication and business management.[27] Gould immediately conveyed Saunders's letter to Osler. "May the devil and his dam take Saunders, and the whole tribe of publishers," Osler exclaimed about this "aggravating affair." Osler again offered to hold some stock himself but advised that Saunders be dropped as a potential publisher: "Do you not think we could influence the Lippincotts?" he asked Gould, "If they had a written guarantee of the support of such men as you could control, and of course the strong backing—from feelings of pride at least—of all the Philadelphia men, they might undertake it. Some years ago a member of the firm was anxious about it, and indeed approached [J. C.] Wilson, who talked to me on the question. Of course they were bitten with the Medical Times (and Register), but tempora mutantur."[28] Gould was rapidly running out of publisher options. After William Wood of New York refused him permission that same year to reprint articles from its medical journals in his *Year-Book*—a digest of material that reached, according to Gould, thousands of readers—he distributed a circular about the relations between the medical profession and "lay publishing firms of medical journals." Publishers do not pay physicians for their contributions, he noted, though they presumably profit from them; and, in this case, no other publisher—even those who do pay contributors—had objected to reprinting extracts. But above all, this publisher's decision was wrong because it prevented the dissemination of medical knowledge. "Is this literature the property of yourself and of the profession or not?" asked Gould. Physician-authors should demand that journals print a statement guaranteeing the right to abstract their text or reproduce their illustrations.[29]

In 1897, Gould published an essay outlining "Some Relations of Author, Publisher, Editor, and Profession" that similarly decried the objectionable state of copyright and unethical practices among "lay" (nonmedical) publishers. In his view, "the right of publication of articles in medical journals is . . . simply of the nature of a temporary loan," and so "if publishers making use of a copyright law, devised for different purposes and conditions, seek to divert the loan to their personal and permanent use and gain, then such publishers are plainly acting in opposition to medical ethics and scientific progress."[30] He continued this line of attack through subsequent publications, in his *Suggestions to Medical Writers*, published in 1900, and reprinted from there in his *Borderland Studies*, published in 1908. In a chapter titled "Some Ethical Questions," his note on "Professional Control, Ownership, and Use of Professional Literature" explained that "when a publisher pays a physician for literary work, the matter is on a different footing; but when the physician gives scientific articles or lends them, without compensation, to be published in a journal, it is plain that these articles are still the property of the author or of the profession, not of the lay publisher."[31] In short, he concluded,

> The lesson to be gathered is that the profession should own and control its journals, and, especially, . . . it should support them rather than journals whose lay owners have always used them for purposes of unadulterated financial

selfishness, and for creating a monopoly of medical literature for which they have not paid the producer a cent.

When he launched a new journal in 1898, the *Philadelphia Medical Journal*, Gould evidently had acted on the advice of his publisher to set up his own company, The Philadelphia Medical Publishing Company, apparently backed by William Osler.[32] As well, he informed readers not only that the journal was "owned, controlled, edited, and published by members of the medical profession" but also that the use "of reading matter and illustrations by reputable physicians, and for legitimate purposes, will not only not be refused, but will be welcomed. It is wholly against the ethics and spirit of the profession to limit the spread of serial scientific literature that has been donated by physicians."[33] He advertised these principles on the journal's masthead: "A Weekly Journal owned and published by The Philadelphia Medical Publishing Company and conducted exclusively in the interests of the Medical Profession." He made a similar statement in the inaugural issue of *American Medicine*, a journal he founded in 1901:

> The publishing company should be incorporated and should have as its principal object the establishment of a professional organ, the profits of which shall go to its purification and perfection, and to progress in general professional concerns. We frankly state that it is our aim to make of American Medicine such an endowed journal. . . . We ask every physician in making his will to consider seriously that the profoundest need of the profession today is a medical journal placed permanently and absolutely beyond commercial temptations and dangers. Nothing else can free us from many abuses that now cripple and disunite us.[34]

In this view, reflecting Osler's wish earlier for an "independent journal," profit should be plowed back into the journal, not into the publisher's pocket.

That Gould would maintain this stance is all the more remarkable under the circumstances: he had been summarily dismissed from the *Philadelphia Medical Journal* by the very publishing company he had organized, for reasons related to commercialism. Undaunted, within a month he had organized and was publishing this new journal, with all the well wishes of colleagues who deemed him "one of the most able and scholarly medical editors on this continent."[35] Typical of his style, his departure from the *Philadelphia Medical Journal* was not without its print record: a protest to the stockholders in the form of a broadside. In it, he provided valuable insights into the running of a medical journal in this period, its management structure and finances, along with the circumstances of his own appointment as editor. His theme emphasized a journal "absolutely untrammeled by commercial or partisan interests," an

The Philadelphia Medical Journal.

A Weekly Journal owned and published by The Philadelphia Medical Publishing Company and conducted exclusively in the interests of the Medical Profession.

The Editorial Staff consists of

GEORGE M. GOULD, M.D., AUGUSTUS A. ESHNER, M.D.,
Editor. *Assistant Editor.*
W. A. NEWMAN DORLAND, M.D., JOSEPH SAILER, M.D.,
ALOYSIUS O. J. KELLY, M.D., C. H. FRAZIER, M.D.,
D. RIESMAN, M.D., D. L. EDSALL, M.D.,
M. B. TINKER, M.D.

Scientific Articles, Clinical Memoranda, News Items, etc., of interest to the profession are solicited for publication. When requested, authors will be furnished 250 reprints of Original Articles.

The Editorial and Business Offices are at 1420 Chestnut St. Address all correspondence to

THE PHILADELPHIA MEDICAL JOURNAL, 1420 Chestnut St., Philadelphia, Pa.
See Advertising Pages 13, 18 and 19.

VOL. I, No. 26.　　　　JUNE 25, 1898.　　　　$3.00 PER ANNUM.

Volume I of the Philadelphia Medical Journal is completed with the present number, including an index of 16 pages. We may be pardoned, we hope, a sense of legitimate pride as we glance over the work done and the success gained. Starting with a promise of 36 pages weekly, we were soon compelled to expand to 44 pages regularly, but even this did not suffice, and we have been under the necessity on several occasions of printing 56. The completed volume for the six months consists of 1226 pages, almost rivaling the work of our long-established and rich English contemporaries, and far exceeding that of any other American medical weekly with the exception of the *Journal of the American Medical Association*—and at a subscription-rate lower than that of all. The volume contains 164 original articles (with 251 illustrations), which will compare favorably with those of the best of our contemporaries, while many have been of exceptional importance and value.

Of appreciation of the fulness and freshness and interest of our News-Departments, supplied by resident medical men in the great centers of the world, we have had most emphatic assurances; and our reports of meetings of Medical Associations and Congresses have been furnished with something approaching the business-ability, and we hope exceeding the accuracy, of the daily newspaper in its sphere of work. Most gratifying letters of praise and expressions of satisfaction for the accounts of the German Medical and Surgical Congresses have been received, and our reports of the Denver meeting of the American Medical Association, secured at great labor and expense, by telegraph, and published the week of the meeting, have astonished and delighted thousands of readers.

To satisfy the popular interest in the medical aspects and lessons of the Spanish-American War we have secured the help of a trained specialist, having exceptionally close relations with the medical and naval departments at Washington, and "at the front," whose letters are serving to correct many errors, to clear up popular comprehension, and to give the latest reliable news in this field. This correspondence will be supplemented by further army and navy news from official and authoritative sources.

Perhaps the most useful and praised portion of our work is the epitomization of the world's latest literature in a thorough and serviceable manner, soon after its appearance in the original, upon a scale and in a manner not equaled by any other journal. To this department alone we have devoted upward of 280 pages. The busy practitioner is thus kept in immediate touch with the discoveries and experiences of the master-minds of medicine without waiting six months or a year for the information, and being spared the expensive and laborious necessity of taking and reading 20 different journals. In a word, we are endeavoring to give the overworked medical man a long-needed tool or professional aid in his busy life. Has it proved useful to you? If so, we frankly ask your sympathy and assistance.

The work has thus far been carried on by real patriotic and professional zeal on the part, first, of a number of noble-minded medical men and laymen who have shown most remarkable unselfishness in a field of thought, on the part of some, alien to ·that of their life; and secondly, on the professional side an enthusiasm and loyalty difficult of appreciation by those not intimately acquainted with the labor involved. Such self-renunciation as has, for instance, been shown by our editorial staff is rarely seen, and we wish to say frankly that it deserves palpable encouragement.

Moreover, the past is only an earnest or promise of our purposes should the profession recognize the opportunity offered. We have by no means realized our ideal. We would like to see medical journalism wholly freed from the stifling influences of commercialism, and of evil lay-motives; we would like to see America lead the world in this as she has in so many other departments of skilled enterprise and democratic progress. Why can we not have a 100-page weekly medical journal serving the whole profession, and promptly reflecting and dispassionately judging the world's progress and discovery? Led by such a voice, what incalculable influence for good could it exert in unifying professional energies, stimulating and controlling sanitary legislation, and commanding therapeutic progress! We are abundantly satisfied with our success in the short time of our young life, but the simple and effective answer to our appeal by each present subscriber is that he feel himself obligated to aid us so far at least as to secure one or two new subscribers. Many have sent us dozens, in recognition of what they have held to be their professional duty. Will you not get one?

Figure 12. Title page of the first volume of *Philadelphia Medical Journal*, edited by Gould from 1898 to 1901.

"independent and scientific professional organ" uncontrolled and uninfluenced by the "domination of publisher and of partisanship"—this latter being the "curse of all medical journalism." He outlined his disgust at those colleagues who had abrogated control of the journal to a general manager who insisted the journal be run like a daily newspaper and rued his "quixotic loyalty" to some rather than accept the editorship of "a great rival journal," which would have provided him a position of "power, influence, and . . . comparative affluence."[36] Clearly Gould was aghast that after all he had written, his own journal had been taken over by someone with not only goals antithetic to the medical profession but declarations that "the capitalist should rule the editorial department."

As time wore on, Gould would expand on his critique of partisan journals along with those run for commercial ends. "And these official medical journals— what a farce they are!" he exclaimed in 1907. "Such journals are carried on for the benefit of the select few who arrogate to themselves a knowledge which has been outlived." In the United States, he argued, the powers of a professional organization were "being used for a most undesirable monopoly, for crushing out democratic spirit and independence, for extinguishing minorities and independent rival journals."[37] His concept of control, then, meant a medical literature separate from professional societies and their organs, such as the British Medical Association, and—presumably in this example—the American Medical Association.

Expanding Collections of Medical Literature

Gould's firm belief in the openness of science underpinned his work toward "public" medical libraries as well. Within months of starting his *Philadelphia Medical Journal* run by physicians, he founded the Association of Medical Librarians. In addition to an exchange of duplicates among its member libraries, Gould envisaged committees to obtain medical literature from targeted sources such as retiring or deceased physicians, society transactions, and auction or antiquarian book dealers. He also sought gifts from publishers. In *Medical News* three years earlier, he had presented his view that publishers would benefit from placing books in libraries. "Some publishers may be so short-sighted as to think that every book that they put into a medical-college library will diminish many-fold that book's sale," he explained; however, "this can only be the case when the book is unworthy. The college-library, in conjunction with the study of medical literature in the college-course, would increase the sale of the best medical literature, and, it is to be hoped, would diminish in some degree the sale of books of low degree now being so persistently exploited upon the medical profession. The systems and text-books of the commercial doctors and cheap publishers are debauching now the taste of the profession."[38]

By 1900, annoyed that "with two noteworthy exceptions" publishers had rebuffed the society's request for gifts because "every book given or even bought by a library reduces their private sales," Gould claimed this was "one result of the commercialism of the medical publishing business. We are convinced that even as a business policy this egregious selfishness is far from being wise and shrewd."[39] He stepped up his criticism of publishers in the next issue of the *Philadelphia Medical Journal*:

> There are not probably half a dozen medical authors who have made or who expect to make a day-laborer's wages for all the time, work, and intellect invested in their books. It is for science and for the profession that they have written. In making contracts, or by special request, they can secure from the publisher 25 or 50 (it should be 50) copies of their books at an inconsiderable expense. Many authors have secured such gifts for the association by renouncing their royalties and securing the volumes from the publishers by gift or at the price of the cost of production. . . . We appeal to every medical author to secure for the membership libraries of the association such gifts of their publications; it would bring them professional honor and gratitude, would not lessen the sales, and would largely help to raise the medical libraries of the country out of their present pitiable and disgraceful conditions of neglect.[40]

Three months later, at the third meeting of the Association of Medical Librarians, Gould was more frank about having "lay and commercial men" in charge of medical literature. He maintained again that sales of books increase when they are available for consultation in libraries.[41] He was quick to exclude his own publisher, Blakiston's and Sons, from his criticism. A gift from his publisher, however, should be seen in light of the enormous success of his dictionaries and his unusual longstanding relationship with Blakiston. Indeed, upon Gould's death in 1922, Kenneth Blakiston remembered their thirty-five year association as "an extremely pleasant one."[42]

This criticism in camera pales next to Gould's published gloat in *American Medicine* three years later. The progress of the Association of Medical Librarians, which had distributed more than five thousand duplicate volumes to libraries in 1902, he pronounced,

> must be gall and wormwood to those firms of medical book and periodical publishers which have derided and discouraged the formation of public medical libraries. . . . medical publishers should be shown the impolicy and stupidity of the selfishness of hating public medical libraries. The greater the number of such libraries the more books will be sold to individuals. Authors of medical books should remember to bargain with their publishers that a copy should be presented to each library on the list of the Association. Professional medical journals of the country give yearly subscriptions to these libraries. Commercial medical journals do *not*. Are you a subscriber or contributor to commercial medical journals?[43]

His question rhetorically accused his readers of supporting commercial publishers' unprofessional activity by subscribing to or publishing in their publications.

Standardizing Medical Language and Communications

Gould did not rest with these efforts to control external matters, the production and dissemination of medical literature; he also worked hard to standardize and simplify medical communications. In a "groping science like medicine . . . naming every step forward is peculiarly necessary," he explained in his supplement *A Dictionary of New Medical Terms*. Quoting Dr. Murray on Johnson's work having raised lexicography to a department of literature, Gould suggested that the technical dictionary "may indeed claim a higher office than that, because . . . more than teacher or text-book [it helps] to bring order into the student's forming mind, and to systematize and make definite his knowledge."[44] Among other accolades for his sustained work in this area, the *British Medical Journal* dubbed Gould "the Johnson of medical lexicography."

Though his own dictionary lasted through half a century, Gould may have been instrumental in the longevity of another. He involved W. A. Newman Dorland in his editorial work, including the *Year-Book* and the *Philadelphia Medical Journal*. Dorland then angered Gould by striking out on his own.[45] Dorland's medical dictionary, first published in 1898, is now the widely used one, still in publication in various media and proudly advertised by its publisher, W. B. Saunders, as "100 Years in Print."

Gould also tried to reform medical language. He sought to abolish "a.m." and "p.m." in clinical records in favor of the twenty-four-hour clock.[46] Where he had control, he exercised it in his publications, as in his vigorous reform of medical orthography. He removed diphthongs, the "u" in "or" endings and reversed "re" to "er" endings—not just simplifying medical language but Americanizing it as well. As he explained, this latter approach was suitably "Yankeelike"—otherwise why not govern, enact laws, and "do everything else on the English models?" Much of his argument for reform was grounded not just in need for consistency in scientific terms but in economy of production: for "the tremendous amount of printing done, and the number of bothered printers, typereaders, and editors in the world, trifles become highly important matters by mere addition."[47]

Although Gould's removal of "ugh" endings in "although," "through," and "thorough" (e.g., "thoroly") renders his publications curious today, these simplifications must be seen in the context of other spelling reforms at the time. His orthography is generally comprehensible in comparison with the more extreme phonetic revisions undertaken by Melvil Dewey, for example. Even so, it drew criticism from some in the medical profession. When its publisher moved the *Medical News* to New York in 1895, an editor of the *Medical Press* happily observed that the "'orthographic atrophy,' which under the former *régime* was so marked a feature of our esteemed contemporary, has now almost disappeared from its pages. Evidently the attempt to foist the new-fangled system of spelling scientific words does not find any favour with the editors of the best organs of the profession in New York."[48]

Despite his uniquely identifiable style—as indicated even in the quotations here—Gould sought a standardized professional voice. Imploring his friend Dr. Bayard Holmes of Chicago to send editorials for *Medical News*, he explained, "I am wretchedly busy, have about written myself out, and feel that I am writing entirely too much for the News myself. My editorials are widely copied, it is true, and highly spoken of, but there should be other minds to form balance wheels and correctives to any individual tendency in editorial work."[49] He did not necessarily see editorials as the work of individuals; rather, he increasingly advocated anonymous editorials and book reviews. Early in 1898 he described the medical journal as a "composite editorial personality" in which the writer should merge his individuality "for the common good." In this way, the profession would avoid the pitfalls of unethical reviews by authors, publishers, and rivals. Signing by full name was not good enough, he argued, for this represented merely one man's opinion; signing by initials—the "neuter-gender review"— was also unacceptable, for then no one appeared fully responsible.[50]

On this occasion, we have evidence of disagreement with his view—and with it, an emerging sense of professionalization in medical editorship. W. A. Young, editor of the *Canadian Journal of Medicine and Surgery*, responded: "We think that the staff of any particular journal . . . should assume the responsibility, if responsibility there be, of their particular views on any subject, otherwise at some juncture there might be a certain amount of confusion." If a coauthor incorporates his own ideas, Young asked, "why should the editor-in-chief have to stand the brunt of the battle if afterwards any discussion ensued on the subject?" If Gould objected to initials as quietly advertising the writer, he was wrong, Young maintained, for "it simply entails a personal responsibility for views expressed."[51]

Others similarly disagreed with Gould's insistence that journal content belonged to all. The editors of the *Montreal Medical Journal* weighed his views on the relations between the medical profession and "lay" publishers of medical journals. As editors, they delighted in having articles from their journal published elsewhere; but as publishers, they saw the potential for objection. Publishers of leading journals have more say than Gould admitted, for a journal that existed to republish articles from high quality journals would likely attract more subscribers than the original journals. They also believed extracts published in yearbooks should be limited. Yearbooks "ought merely to publish extracts sufficiently full and sufficiently suggestive to lead those who read them to consult the original articles. There is in this country altogether too great a tendency to subsist upon second-hand matter, and an unwillingness to consult original authorities." These Canadian editors concluded that if Gould wished to publish only abstracts in his yearbook, then they would censure William Wood & Co. for not giving permission; however, if he wished to publish large verbatim portions either out of context or without payment to authors, then they would side with the publisher.[52]

Just as he had earlier heeded the advice of friend and publisher in establishing the *Philadelphia Medical Journal*, so too did Gould revise his policies based on such reactions to them. In reviewing the newest column in the journal in December 1898, one that abstracted articles from other journals, he announced: "The collaborators who do the work of reading and condensing the original articles of the best medical journals are scholarly and devoted men, whose labor is enormous, and who are doing the profession a high and unselfish service deserving of all honor. We shall hereafter append to each abstract the initials of the collaborator."[53]

George M. Gould and the Business of Writing Medicine

As this discussion suggests, whether engaged in writing or righting medicine, Dr. George M. Gould magnifies medical print culture in the Progressive Era. His unique activities not only embrace all genres in medical publishing—dictionaries, digests, treatises, pamphlets, journals, and histories—but they also provide a leader's view of their value and use in shaping medical knowledge. He established a central exchange of medical publications that continues to the present, along with its parent society that has continued his goal of improving medical library collections. Although some editors disagreed with his attempts to standardize medical language and professional communications, many leaders in medicine, such as William Osler and Alfred Stillé, supported his insistence that the profession control its own publications.

His work spotlights the journal as a large part of the publishing enterprise—a part often overlooked in favor of the book in institutional histories of medical publishers.[54] More important, it places him as forerunner on a continuum of outspoken medical editors in the twentieth century. Through three Philadelphia-based national journals he directed until 1906, Gould became known internationally for a vigorous editorial style that aimed always "to correct some abuse" in medicine. "It is true that Dr. Gould had in some instances taken somewhat extreme views on certain points," observed W. A. Young, but as he was "'the man at the front,' and the 'man with the gun,'" no one would say that "he has not published a medical journal that has been a credit to him as a writer and to the profession of whom he is so prominent and highly esteemed a member."[55] As well, that Gould was based in Philadelphia speaks to the centrality of this city in North American medical publishing before the end of the nineteenth century.[56] Indeed, *Medical News*'s move to New York in 1895 was cause for concern among observers who clung to the view that Philadelphia was the "medical Mecca of North America."[57] For this reason, Osler urged Gould to found an independent journal for "the old town," with the proud backing of "all the Philadelphia men."

Both Gould's publications and private correspondence provide insights into the evolving relationship between publisher and medical profession. He railed against publishers who owned medical journals or underwrote medical books for refusing to make this literature available to the medical profession without cost; medical professionals write for these journals without payment for their contributions, he explained, yet the publishers reaped the profits. Publishers disagreed with him because they underwrote the cost of printing and distributing the journals; and whether or not the medical profession entirely agreed, he did reflect its desire to control everything pertaining to medical education and practice. His stature and this push for medical control had its downside, however, for it meant Gould was fired from his post—by his medical colleagues at the *Philadelphia Medical Journal*. Although these men might characterize as arbitrary Lea's decision to move a journal from Philadelphia to New York, when all is said and done, it was not the nonmedical publisher who dismissed its editor. Moreover, it had not seemed to occur to Gould that control by colleagues could be harder than control by a publisher: medical practitioners would be less tolerant of editorial deviation from views about how to conduct *their* journal; they had less invested than a publisher; and they acted voluntarily as managers. Colleagues could therefore be maneuvered by their nonmedical employees into dismissing the editor (a scenario that would repeat in dismissals of later North American medical editors).

Nevertheless, in all his publishing endeavors George Gould's struggle with a single question throws into relief a perennial issue for the medical profession: who owned the information being generated by its practitioners? As those active in science and medicine know, the issue of ownership is still hotly debated by researchers, publishers, editors, and observers who fear the monopolization and commodification of knowledge.[58] Richard Smith has explored some concerns in a section of his book entitled "Important Relationships of Medical Journals": the development of embargoes from journals on information for the mass media; journals and pharmaceutical companies as "uneasy bedfellows"; and "the highly profitable but perhaps unethical business of publishing medical research." In the latter chapter, he poses as provocative a series of questions as would make Gould proud:

> When authors submit a paper to a journal they must usually agree both not to submit elsewhere and to transfer copyright (for no fee). Why, I often wonder, do authors agree to these requirements? . . . There is intense competition for the

best papers. Why don't authors announce that they have an excellent paper and then ask journals to bid for them? Various authors have thought of this idea, but it requires concerted action to break the stranglehold of the journals. And why do authors hand over copyright? Some publicly funded organizations— like Britain's Medical Research Council—do not, and the journals have to accept it. When they do hand over copyright authors have to ask permission from publishers in order to reproduce their own material . . . an extraordinary state of affairs, and increasingly journals do allow authors to do what they want with their own material—including placing it on their own or their institution's website.[59]

Gould's organizational efforts in the print medium thus join a succession of enterprises that seek to place medical and scientific knowledge into the realm of open access—at least for professionals. He no doubt would applaud the arrival of the Public Library of Science, Creative Commons licensing, and BioMed Central.[60]

Finally, a key concept cohering all this activity in medical publishing, from Gould's time to the present, is "ethics." Gould and his contemporaries cast their concern over control in terms of ethical practices, just as Smith and his editorial colleagues have done today. Titles of their essays—from Foshay's "Medical Ethics and Medical Journals," Gould's "The Ethics and Politics of Medical Book-reviews," Young's "The Ethics of Medical Writing," and Gould's chapter "Some Ethical Questions"—mirror those by Smith and Jones and McClellan, perhaps culminating in Smith's "Ethical Manifestos for Four Different Futures for Medical Publishing." Fuller consideration of these "ethics" is beyond the scope of this discussion, but concluding with another example of Gould's outrage suggests both his command over the subject and his place in its history. In 1907, Gould observed that "whole articles and books and 'systems' exist, not a page of which was written, and often not read, by the men credited with the authorship. One recent pompous article in a big book on an immensely common nervous disease was old medieval stuff recooked by a penny-a-liner."[61] His intent was to implicate medical publishing in attracting the unscrupulous to the profession. Although he referred to a derivative essay in an obscure book that had been written for payment, his raising the alarm aligns him with like-minded physicians and medical editors who criticize publishing developments since World War II, with the rise of "big science" and the pharmaceutical industry. Their studies, unlike those of other disciplines in book and print history, share Gould's urgent desire to "right" medicine by uncovering abuses in its publication.[62]

Notes

1. P. Maxwell Foshay, "Medical Ethics and Medical Journals," *Journal of the American Medical Association* (hereafter *JAMA*) 34 (April 28, 1900): 1041–43.

2. Richard Smith, *The Trouble with Medical Journals* (London: Royal Society of Medicine Press, 2006), 266.

3. Anne Hudson Jones and Faith McLellan, eds., *Ethical Issues in Biomedical Publication* (Baltimore: Johns Hopkins University Press, 2000).

4. International Committee of Medical Journal Editors, "Uniform Requirements for Manuscripts Submitted to Biomedical Journals," available at http://www.icmje.org.

5. See my own article, "Publisher Ownership, Physician Management: Canadian Medical Journals in the Victorian Era," *Victorian Periodicals Review* 34, no. 4 (2010): 388–428. Lack of attention to medicine and print culture has otherwise meant a reliance on studies of science, which often subsumes medicine as a field within it. See, for example, Alan G. Gross, Joseph E. Harmon, and Michael Reidy, *Communicating Science: The Scientific Article from the 17th Century to the Present* (New York: Oxford University Press, 2002); and Mario Biagioli and Peter Galison, eds., *Scientific Authorship: Credit and Intellectual Property in Science* (New York: Routledge, 2003).

6. The most important essay collection remains W. F. Bynum, Stephen Lock, and Roy Porter, eds., *Medical Journals and Medical Knowledge: Historical Essays* (London: Routledge, 1992). Other collections that include books and libraries and are more grounded in the approaches of book history include Jennifer J. Connor, ed., *Book Culture and Medicine*, in *Canadian Bulletin of Medical History* Special Issue 12, no. 2 (1995): 203–445; Robin Myers and Michael Harris, eds., *Medicine, Mortality and the Book Trade* (New Castle, Del.: Oak Knoll, 1998); and Charles E. Rosenberg, ed., *Right Living: An Anglo-American Tradition of Self-Help Medicine and Hygiene* (Baltimore: Johns Hopkins University Press, 2003).

7. George M. Gould, "Medical Journalism and Medical Libraries," *Philadelphia Medical Journal* 1, no. 24 (June 11, 1898): 1071–72.

8. "The Association of Medical Librarians," *Philadelphia Medical Journal* 1, no. 26 (June 25, 1898): 1170.

9. Except where noted, the following capsule biography of George Milbry Gould (1848–1922) is derived from chapter 2 of Jennifer Connor, *Guardians of Medical Knowledge: The Genesis of the Medical Library Association* (Lanham, Md.: Scarecrow Press, 2000), 41–43. Full citations to sources are provided there.

10. Reviews of George M. Gould, *Borderland Studies* in College of Physicians of Philadelphia, George Gould Scrapbooks, 10c 56, vol. 57:66, 80, 87 (hereafter CPP, Gould Scrapbooks).

11. "Book Notices," *JAMA* 28 (January 9, 1897): 92.

12. Lafcadio Hearn to George Gould, 1889, in Elizabeth Bisland, *The Life and Letters of Lafcadio Hearn*, 2 vols. (Boston: Houghton, Mifflin, 1906), 1:464. His own *Bibliography of the Contributions of George M. Gould, M.D. to Ophthalmology, General Medicine, Literature, etc.* (Ithaca, N.Y.: Andrus & Church, 1909) lists well over four hundred items.

13. Jefferson graduation program, CPP, Gould Scrapbooks, vol. 68:29.

14. Except where noted, this discussion of the Medical Library Association Exchange is derived from Connor, *Guardians of Medical Knowledge*, chap. 2.

15. "The Organization and Support of Medical Libraries," *JAMA* 29 (October 9, 1897): 748; also published as "The Organization and Support of Public Medical Libraries," *Bulletin of the American Academy of Medicine* 3, no. 2 (1897): 148–49.

16. For fuller discussion, see Connor, *Guardians of Medical Knowledge*, 64–65.

17. "Report of the Meeting of Medical Librarians, Held in Denver June 6, 1898," *Medical Libraries* 1, nos. 5–6 (June-July 1898): 36–39.

18. "Association of Medical Librarians: Minutes of First Meeting," reprinted in Albert Tracy Huntington, "The Association of Medical Librarians: Past, Present, and Future," *Medical Library and Historical Journal* 5 (1907): 111–23, esp. 114.

19. Jennifer J. Connor, "Only for 'Purely Scientific' Institutions: The Medical Library Association's Exchange, 1898–1950s," *Journal of the Medical Library Association* 99 (2011): 118–26.

20. Ralph Elmergreen to George Gould, 6 January 1896, CPP, Gould Scrapbooks, vol. 57:105.

21. Alfred Stillé to George Gould, 8 November 1895, CPP, Gould Scrapbooks, vol. 87:70.

22. Letters and editorials, CPP, Gould Scrapbooks, vol. 57:96, 115, 106, 113.

23. Wally Reynolds, "The Most Brilliant Medical Editor America Has Produced," *American Medicine* n.s. 17, no. 28 (September 1922): 524–25.

24. William Osler to George Gould, n.d. [1895], CPP, Gould Scrapbooks, vol. 87:47.

25. C. M. Lea to George Gould, 6 February 1892, CPP, Gould Scrapbooks, vol. 68:58.

26. Charles M. Lea to George Gould, 24 January 1893, CPP, Gould Scrapbooks, vol. 68:62.

27. William Saunders to George Gould, 22 June 1896, CPP, Gould Scrapbooks, vol. 68.

28. William Osler to George Gould, 30 June 1896, CPP, Gould Scrapbooks, vol. 68.

29. "Medical Writers and Publishers," *Montreal Medical Journal* 25 (January 1897): 595–97.

30. George M. Gould, "Some Relations of Author, Publisher, Editor, and Profession," *Bulletin of the American Academy of Medicine* 3, no. 2 (1897): 110.

31. George M. Gould, *Suggestions to Medical Writers* (Philadelphia: Philadelphia Medical Publishing Company, 1900), 143–45; and Gould, *Borderland Studies*, 2 vols. (Philadelphia: P. Blakiston's Son, 1908), 2:209–11.

32. Harvey Cushing, *The Life of Sir William Osler*, 2 vols. (Oxford: Clarendon, 1925), 1:484.

33. "'Copyrighted,'" *Philadelphia Medical Journal* 1 (January 1, 1898): [1].

34. "To Encourage the Endowment of Medical Journals," *American Medicine* 1 (April 20, 1901): 98.

35. W.A.Y. [W. A. Young], "Dr. Geo. M. Gould's Temporary Retirement from Medical Journalism," *Canadian Journal of Medicine and Surgery* 9, no. 2 (1901): 130.

36. "A Protest Made by Dr. George M. Gould, at the Stockholders' Meeting of the Philad'a Medical Publishing Co., Held January 8th, 1901," CPP, Flick Pamphlets, vol. 19, pamphlet 42. Within a year of starting the *Philadelphia Medical Journal*, Gould apparently was sought as editor of the *Journal of the American Medical Association*. Osler observed that *JAMA* needed him more, for he could place it "on a first class basis": see Osler letter to John Musser, 31 December 1898, quoted in Cushing, *Life of Sir William Osler*, 1:484–85.

37. Gould, "Vocation or Avocation?" in *Borderland Studies*, 2:304–5.

38. "The Influence of Medical Libraries on Medical Literature," *Medical News* 67 (August 24, 1895): 217.

39. "The Association of Medical Librarians," *Philadelphia Medical Journal* 5 (March 24, 1900): [645].

40. "To Disseminate Medical Literature and Encourage Public Medical Libraries," *Philadelphia Medical Journal* 5 (March 31, 1900): [697].

41. Report of George Gould, reprinted in Huntington, "Association of Medical Librarians," 118–19.

42. Kenneth M. Blakiston, "The Johnson of Medical Lexicography," *American Medicine* n.s. 17, no. 28 (September 1922): 525–26.

43. "The Association of Medical Librarians," *American Medicine* 5 (January 10, 1903): 43.

44. George M. Gould, "Notes Concerning the History of Lexicography," in *Gould's Medical Dictionary*, 2nd ed., ed. R. J. E. Scott (Philadelphia: P. Blakiston's Son & Co., 1928), xi–xiii.

45. W. A. Newman Dorland and George Gould letters, 1898, CPP, Gould Scrapbooks, vol. 69.

46. "24 O' the Clock," *Philadelphia Medical Journal* 1 (January 22, 1898): [169].

47. Gould, *Suggestions to Medical Writers*, 40–41.

48. Clipping, *Medical Press*, 15 February 1896, CPP, Gould Scrapbooks, vol. 57:91.

49. George Gould to [Bayard] Holmes, 26 September 1893, George Gould Papers, MS C 138, box 1, file 1893 1–16, History of Medicine Division, National Library of Medicine, Bethesda, MD. This lament was not unusual, for he wrote years earlier, "Other literary and medical jobs are being fired at me, till I wonder if I shall get time to eat lunch by and by. . . . I s'pose there'll be eternity to rest in before long." George Gould to James Bonar, 24 May 1887, box 1, file 1886–92 1–15.

50. "The Ethics and Politics of Medical Book-reviews," *Philadelphia Medical Journal* 1 (March 19, 1898): 481.

51. W.A.Y. [W. A. Young], "The Ethics of Medical Writing," *Canadian Journal of Medicine and Surgery* 3, no. 4 (1898): 222.

52. "Medical Writers and Publishers." For fuller discussion, see Bertrum H. MacDonald and Jennifer J. Connor, "Science, Technology, and Medicine: Constructing Authorship," in *History of the Book in Canada*, vol. 2, *1840–1918*, ed. Yvan Lamonde, Patricia L. Fleming, and Fiona A. Black (Toronto: University of Toronto Press, 2005), 177–86.

53. "Our Latest Literature Department," *Philadelphia Medical Journal* 2 (December 3, 1898): 1150.

54. For example, R. Kenneth Bussy, *Two Hundred Years of Publishing: A History of the Oldest Publishing Company in the United States, Lea & Febiger, 1785–1985* (Philadelphia: Lea & Febiger, 1985); and John L. Dusseau, *An Informal History of W. B. Saunders Company* (Philadelphia: W. B. Saunders/Harcourt Brace Jovanovich, 1988).

55. W.A.Y., "Dr. Geo. M. Gould's Temporary Retirement."

56. For more information, see Jennifer J. Connor, "Stalwart Giants: Medical Cosmopolitanism, Canadian Authorship, and American Publishers," *Book History* 12 (2009): 209–39.

57. *Times and Register* item reprinted in *Lancet Clinic*, 26 January 1896; and *North American Practitioner*, February 1896, CPP, Gould Scrapbooks, vol. 57:90, 97.

58. See, for example, discussions posted on the *Forum on Privatization and Public Domain*, available at http://www.forumonpublicdomain.ca.

59. Smith, *Trouble with Medical Journals*, 212.

60. For a case study, see John Willinsky et al., "Doing Medical Journals Differently: Open Medicine, Open Access, and Academic Freedom," *Canadian Journal of Communication* 32, no. 3 (2007): 595–612.

61. Gould, "Vocation or Avocation?" 302–3.

62. See Jeremy A. Greene and Scott H. Podolsky, "Keeping Modern in Medicine: Pharmaceutical Promotion and Physician Education in Postwar America," *Bulletin of the History of Medicine* 83 (2009): 331–77, esp. 371–75. For a recent case study, see Joseph S. Ross et al., "Guest Authorship and Ghostwriting in Publications Related to Rofecoxib," *JAMA* 299, no. 15 (2008): 1800–1812.

Science Education and
Health Activism in Print

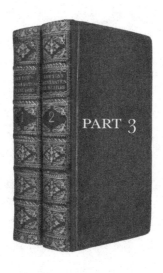

PART 3

Evolution in Children's Science Books, 1882–1922

KATE MCDOWELL

In 1922, Hendrik Van Loon received the first Newbery Medal, for his book *The Story of Mankind*, a tour of humankind through the ages, beginning with the evolution of human beings. The Children's Librarian's Section of the American Library Association created the Newbery Medal to be given to the best book published in the United States for children.[1] In the history of children's literature, this first Newbery Medal winner is an anomaly; few nonfiction works have won the medal, and fewer still works of science have received Newbery medals or recognition as Newbery honor books. In fact, there have only been two others: a 1940 honor-winning biography, *Runner of the Mountain Tops: The Life of Louis Agassiz* by Mabel Robinson, and a 1956 honor-winning introduction to biology, titled *Men, Microscopes, and Living Things by* Katherine Shippen. Although Van Loon's work is unusual when compared with what has won the Newbery award since, it was continuous with a strong emphasis on natural history and nature study in late-nineteenth- and early-twentieth-century publishing for children.

However, Van Loon's book was still unusual among natural history works for children in that it addressed a controversial issue; it opened with a chapter addressing the most controversial aspect of evolutionary theory, human evolution. Van Loon introduced an account of evolution that traced the "ascent of man" from single-celled life-forms in the sea to "Early Man." Van Loon implied support for Darwin's theory of natural selection when he wrote: "Only the people with the cleverest brains survived."[2] While there were many science

books published for children in the United States before 1922, relatively few
of them described or discussed evolution, and fewer still explained human
evolution.

To discover how often evolution appeared and how it was described to
young readers, this chapter analyzes science trade books written for young
people and recommended from 1882 to 1922. Evolution was explained as natural
selection, alluded to through adaptation, and, occasionally, used to connect
humankind to the larger animal kingdom. Children's science books typically
referenced evolution by describing concrete, observable relationships between
creatures such as insects and flowers. The ways evolution appeared and did not
appear in these texts raises questions about the degree to which most authors
were able to convey it as a concept, with its vast span of time and complex
relationships among organisms, to a child audience.

Despite this historical emphasis, works of science written for and recommended to young people have rarely been the subject of literary scholarship;
children's literature scholarship tends to investigate children's fiction, with a few
exceptions.[3] Similarly, historians of science have given relatively little attention
to children's science education, including the books or textbooks used, available,
or recommended.[4] Educational scholars have made a few studies of evolution
in high school biology textbooks, but such an approach elides both science
trade books and what younger children could have chosen for themselves in
public libraries.[5] How evolutionary theory as expressed in Charles Darwin's
ideas was disseminated to various groups of people has been the subject of some
scholarship. However, investigated groups are generally presumed to be adults,
organized by their geography, gender, race, and religion.[6] Children are both
part of each of these social groups and distinct from adults in them, but their
vantage point has, for the most part, been overlooked in these investigations.

The period from 1882 to 1922 not only spans the time from librarians' early
recommendation lists to the inauguration of the Newbery medal, it is also
significant because it preceded the famous Scopes evolution trial in 1925, which
marked a period of escalating cultural conflict over teaching evolution. The
appearance of evolution in materials published for and recommended to
children during this period provides a window into how broad cultural interest
in evolution filtered into children's collections prior to the Scopes trial. Librarians
in the late nineteenth century saw the emergence of professional librarianship
and the specialization of services to children, which provides a lens for seeing
how ideas about evolution were distributed to children.[7] However, little research
has yet been done on the market for children's science books during this period,
evidence of their readership, or analysis of their critical reception, which raises
a number of questions beyond the scope of this essay. For example, almost half
of the recommended books were first published in Britain, and though some
scholars have examined differences between British and American children's

fiction, science books have not yet been examined comparatively.[8] The lists were specific to the American context, so this analysis is of what might have been available to children in the United States in public library collections.

Librarians began creating recommendation lists for young people in 1882, when Caroline Hewins of the Hartford Public Library published her influential list, *Books for the Young*.[9] Hewins was author or contributing author for several of these lists, so her influence in these recommendations was potent, as it was nationally in the emerging specialization of children's librarianship.[10] The books recommended in this and subsequent lists were intended to help new professionals in the emerging area of children's services to efficiently select books for their libraries. These early children's librarians were the most authoritative experts in the selection of children's materials during this period and were acknowledged as "the first to put in operation a practical plan for bringing the best books within reach . . . of children" by their colleagues at the National Education Association (NEA).[11] Between 1882 and 1922, librarians selected science books alongside all other sorts of fiction and nonfiction for these lists, which were used to purchase books for public library shelves, school library collections, and in the case of Hewins's lists, the prefaces indicate that they were intended for parents and children as well.[12]

In the early part of this period, many children's librarians recommended natural history, history, travel, and biography over fiction, so analyzing science books for their evolution content sheds greater light on those nonfiction titles recommended as the best reading for youth.[13] That Hewins's first list had an immediate impact on the field was evidenced in quotations from librarians who called her list a "handbook" full of "wisdom" in its selections.[14] Hewins continued to publish highly selective lists of what she considered the best in children's books under the title *The American Library Association's Annotated List of Books for Boys and Girls*, with new editions in 1897, 1904, and 1915. The American Library Association (ALA) also commissioned John Sargent and his sisters Mary E. Sargent and Abby L. Sargent to compile a larger list of recommended titles; they incorporated Hewins's recommendations but expanded upon her list by adding newer titles in their 1890 list and 1896 supplement.[15] The ALA also published *Catalog of ALA Library: 5000 Volumes for a Popular Library* in 1893, and this activity culminated in the 1909 publication of the *Children's Catalog* from H. W. Wilson, which thereafter became the standard list for children's librarians' collection development work, with new editions published in 1916, 1917, and 1925.[16]

This analysis draws on the science book recommendations made in ten lists, published from 1882 to 1925, each of which was influential in determining what public libraries would collect for children. As author of or contributor to many of the lists, Hewins was unusually influential, but this reflects her status as the central national figure in the development of children's librarianship in the

Table 1:   Ten Lists of Recommended Books, Published from 1882 to 1925

1882 *Books for the Young*, Caroline M. Hewins

1890 *Reading for the Young: A Classified and Annotated Catalogue with an Alphabetical Author-Index*, John Frederick Sargent

1893 *Catalog of ALA Library: 5000 Volumes for a Popular Library*

1896 *Supplement to Reading for the Young: A Classified and Annotated Catalogue with an Alphabetical Author-Index*, John Frederick Sargent

1897 *Books for Boys and Girls; A Selected List*, Caroline M. Hewins

1904 *Books for Boys and Girls; A Selected List*, Caroline M. Hewins

1909 *Children's Catalog*

1915 *Books for Boys and Girls; A Selected List*, Caroline M. Hewins

1917 *Children's Catalog of Thirty-Five Hundred Books: A Guide to the Best Reading for Boys and Girls*, Corinne Bacon

1925 *Children's Catalog: A Dictionary Catalog of 4100 Books with Analytical Entries for 863 Books*, Minnie Earl Sears

---

1880s. Statistical prominence in the lists should not be read as absolute cultural prominence, and yet these lists are the only reasonably authoritative national sources for understanding what was likely to have been purchased in public libraries for children during this time period. Most of the lists include publishers' prices; the books ranged in price from a low of $0.60 for Sophie Herrick's 1888 *The Earth in Past Ages* to a high of $3.50 in 1914 for William T. Hornaday's *American Natural History*.[17]

The lists were typically organized by topic, ranging from the nonstandard and sometimes quirky topics created by their individual authors to later use of Dewey decimal classification in various editions of *Children's Catalog*. Catalogers know subject headings change over time. In the case of these lists, subjects could change from edition to edition, and the same book was often classified in different ways from year to year, even when the same authors generated the recommendation lists. Patterns and suggestive changes as to how each text was classified are also discussed as relevant to the context of the recommendations. Thirty-three books featuring some aspect of evolution appeared in these lists under topics as diverse as "out of door books, " "natural history," "prehistoric man," and "cave dwellers" (see appendix). At least one of these lists placed all of the books analyzed here under recognizable science categories.

Digitized texts provided the source material for much of the following analysis. Because of their publication dates, the books themselves are in the public domain, and many are available online in full, searchable text editions, via services such as the Internet Archive, Project Gutenberg, and Google

Books. These large-scale digitization projects made it possible to compare texts with greater efficiency than ever before. Keyword searches for "Darwin," "evolution," "natural selection," "adaptation" and other words provided a first indicator of whether evolution appeared, which was then followed by in-depth reading and examination of the texts.[18] Previously, such a large-scale comparison of the contents of children's texts would have been difficult, to say the least, requiring travel to multiple archives, and the keyword searches made possible by digitization greatly enhanced the efficiency of both data collection and analysis. Nevertheless, some texts were available only in print; For these, found at the University of Illinois libraries at Urbana–Champaign and at the Newberry Library in Chicago, searches were conducted using indexes, tables of contents, and examination of potentially relevant chapters.

Most of the searches produced books that mentioned, described, or explained evolution, but several books were eliminated when the authors took antievolution stances. A book by Francis T. (Frank) Buckland, the 1860 *Curiosities of Natural History*, did feature the word "adaptation" in ways that seemed to suggest evolution but was eliminated from the list because of the author's anti-evolution stance in the 1891 *Log-Book of a Fisherman and Zoologist*. Buckland's fictional monkey narrator states that she is "pleased to hear, from Mr. Darwin's green book, that he is descended from some of my ancestors, and that he is a near relation of mine," followed by: "I see the reader is laughing at my letter."[19] Buckland frames the reader as laughing at the assertion that evolution is a plausible theory.

As natural selection was Darwin's central contribution to evolutionary theory, those texts that mention either the man or his ideas demonstrated that Darwin's popular acclaim did, to some degree, filter into children's literature. In other texts, "evolution" or "adaptation" appeared without indication of whether these words referred to Lamarckian or Darwinian evolution. The former, based on the ideas of Jean-Baptiste Lamarck, refers to the idea that children could inherit traits from the acquired characteristics of their parents. In contrast, the latter rests on the more widely accepted view that adaptation occurs randomly and constitutes a survival advantage only when traits happen to be beneficial to an organism, with evolution operating by this process of natural selection among these randomly generated traits.

Even allowing for duplication, more than 250 science trade books for children received recommendation in at least one of the ten lists published from 1882 to 1925. Thirty-three of these contained some evidence of the cultural impact of evolution, ranging from explaining Darwinian natural selection to implying evolution with words such as "adaptation." Even before controversies over science education sparked by the Scopes trial, there was a relative paucity of evolution in children's science trade books.

Nature and Children's Print Culture

During the Progressive Era, as more institutions emerged with programs to educate and protect children, taking children, especially urban children, to the natural world was increasingly valued. Boy Scouts, Campfire Girls, other nature clubs and camps, outdoor sports, and general forays into the wild became standard childhood fare. Playgrounds became a means for urban children to safely experience some of the benefits of nature.[20] When children learned about the natural world in school, they were taught to engage in "nature study," science inquiry that focused on their direct encounters with nature.[21] Nature study approaches that gave both observation and collection instructions for different specimens were echoed in biology books for children. To give one literary example, Louise May Alcott's book *Little Men* featured a section in which "Jo's boys" created a "museum" of their own out of "the curious and interesting things" they found outside, including pebbles, mosses, and butter-flies.[22] Common terms for such collections included "cabinets of curiosities" and "schoolboy museums."

There were some features of the cultural context around children's print culture that make books of science from this time rich for investigation. During the 1880s and early 1890s, many librarians balked at children's "craze" for fiction reading and instead recommended nonfiction, promoting books of science and natural history above all else as wholesome reading for the young.[23] During the Progressive Era, beginning approximately in the 1890s, children's hands-on experiences through nature study received attention as philosophically connected to Progressive educational ideals, such as John Dewey's ideas of inquiry and learning as experience.[24] Nature study both capitalized on and expanded the cultural association between children's science education and the experience of the direct observation of nature. Some even claimed that nature study taught not only facts about the world but also an "outlook," or a way to relate to the world.[25]

Late-nineteenth- and early-twentieth-century publishing for children saw much emphasis on the natural world and the idea of nature study in particular.[26] Typical books included topics such as animal behavior or floral displays. Books for children used the structure of fairytales to humanize their science stories or relied on the narrative devices such as a talk by the hearth or a walk by the seaside. Children's books, like adult books, often depicted the naturalist as adventurer or explorer.[27] Science authors writing for children used fabricated dialogues between anthropomorphized animals, plants, or even natural forces such as the sun to help young readers see themselves in relationship to nature. Their texts moved from the objective scientific approach toward "specimens" to the close emotional relationship between a child and a pet. Some writings combined observations of nature with musings on God's creation in a form of natural theology designed to appeal to their popular audience.[28]

One unusual children's science author from this era, Arabella Buckley, has been the subject of historical scholarship. For many years, she worked as assistant to Charles Lyell, mentor to Charles Darwin. She therefore occupied an unusual position in the scientific world well before she turned her attention to writing for children. Historian Barbara Gates emphasized the sophistication of Buckley's writings for children in that they moved young readers from what could be observed to complex scientific ideas that eluded observation. Evolution was at the core of Buckley's scientific thinking, and, as Gates argues, her works for children went so far as to imply the evolution of human morality as indicative of the progress of the species.[29] Several of her works, which were repeatedly recommended, are discussed below.

In examining science books from this period, making a precise distinction between those written for children and those written for adults can be difficult. That books of science were intended for a child audience is sometimes stated in titles, as with *Children's Stories of the Great Scientists* by Henrietta Christian Wright, or in subtitles, such as *Madam How and Lady Why: First Lessons in Earth Lore for Children* by Charles Kingsley. In other cases, the intended audience is made clear in the preface. It must be noted that childhood was an emerging cultural concept during this period, especially prior to the widespread adoption of such social regulations as anti–child-labor laws and mandatory school attendance. The fixing of age-level distinctions came about slowly after 1850, as more public schools began dividing children into classes based on their years of birth. In the late nineteenth century, childhood was characterized more by "dependency and innocence" than particular age ranges. Some ten-year-old factory workers functioned culturally as adults, while middle- and upper-class youths of seventeen remained firmly ensconced in the protection of childhood.[30]

Distinctions between children and adults are relevant because there were quite a few books written for adults that librarians recommended for children; scholars of the history of science will be familiar with authors such as Thomas Huxley, Frances Galton, Ernst Haeckel, John Lubbock, and others who wrote such works. However, to focus more closely on how the theory of evolution was written for and disseminated to child audiences, those works clearly written for adults have been excluded from this analysis Additionally, science books written for adult audiences were recommended less frequently in the lists than books written explicitly for children. While many books written for children were recommended in two or more of the nine lists, popular science books written for adults were rarely recommended in more than one list.

Darwin and Evolution in Books for Children

All together, thirty-two science titles recommended for children from 1882 to 1925 contained some positive reference to evolution. Of those, only two explained

Darwin's theory of natural selection. Other books alluded to evolution indirectly by describing some observable form of insect, plant, or animal adaptation. A handful of books published near the end of the period, especially in the early 1920s, included human evolution alongside animal and plant evolution.

There was some evidence of Charles Darwin's influence in these books. For instance, in its concluding "Notice to Teachers," Edward Sylvester Morse's *First Book of Zoology* referenced Darwin's writings among other recommended authors, books, and periodicals for further study.[31] An abbreviated biography of Darwin appeared in a chapter on "Darwin and Huxley" in *Children's Stories of Great Scientists* by Henrietta Christian Wright.[32] Some references were even more oblique, as in Arabella Buckley's *Life and Her Children,* in which an image of an acorn barnacle is credited to Darwin, although no other mention of Darwin's works or theories appears in the text.[33] However, another work by Buckley, *A Short History of Natural Science,* provided a sweeping look at scientific thought through the ages and included Darwin with Lyell, Agassiz, and others in her chapter on "Science in the Nineteenth Century." She also included descriptions of both natural selection and adaptation in concrete terms, as discussed below.[34]

In another example, Spencer Trotter's *Lessons in the New Geography* mentioned Darwin multiple times as a scientist whose work had inspired interest in the connections between natural creatures and their location. In explaining the new significance of geography, Trotter wrote: "*Time* and *Place* became the fulcra on which thought was levered in turning the great questions of life. Evolution was the word and the light of science" (emphasis in original). He argued that similarities and differences among creatures in different locations served to expand the idea of place in understanding natural history and that Darwin's observations on that trip "opened out a broad and suggestive field of thought."[35]

Charles Darwin's own book adapted for children, *What Mr. Darwin Saw in His Voyage Round the World in the Ship "Beagle"* was by far the most frequently recommended children's science book that had any connection to evolution. It appeared in six of the ten lists and was typically categorized with travel and adventure stories, although one list classed it with zoology and another with natural history. The book included separate prefaces for parents and children. In the preface for parents, the stated purpose of the book was "to interest children in the study of natural history, and of physical and political geography." The preface for children took a different tone, encouraging them to look at nature closely, carefully, and to develop an educated eye, because "those see best who know the most, or who naturally take notice of new things." The book did not explain evolution but instead recounted Darwin's observations and urged children to make their own, similar observations. The preface concluded with an activity in which children are invited to "try how good a seer you are by counting the various animals shown in the wood-engraving on the opposite page."[36]

Natural Selection Explained

In only a few texts was the theory of natural selection explained at a child's level of comprehension. Two books both named the theory and illustrated it with some concrete examples. In *A Short History of Natural Science*, Buckley explained the process of natural selection that took place in the development of certain kinds of birds' wings:

> Then those descendants of the strong winged bird which also have strong wings will be most likely to live on in each generation and will pass on this peculiarity to their children; while the descendants of the dark coloured bird will also survive in each generation exactly in proportion as their plumage is adapted to hide them; and thus the strong winged birds and the dark winged birds will in time become very different from each other. This is roughly the theory of "Natural Selection;" that nature allows only those animals to live which in some way escape the dangers which threaten their neighbours, and thus in time the race becomes altered to suit the life it has to lead.

Buckley's emphasis is on the winners, or the animals that live on or escape, rather than those who lose the struggle for survival. It is notable that she details not only the process of natural selection for certain characteristics but also the divergence of different kinds of birds from a common ancestor. Later in the text, Buckley writes that animals all came from one "common source" and that how they "adapted to changed habits and conditions of life" accounts for their great variety, a statement that could imply a Lamarckian view of evolution if she had not earlier explained natural selection.[37]

Similarly, in *Real Things in Nature*, author Edward Holden features a section titled "Natural Selection—The Struggle for Existence" that explained natural selection with reference to the camouflage coloring of deer:

> Every animal gets its food and saves its life from enemies by a struggle for existence. The fittest survive, the weaker die. It is the same with plants and trees; with fish and birds. Why do you suppose, are most wild animals, deer for instance, of the same color on both sides? And why is that color the color of the regions in which the deer live? Because a dun colored deer is not so easily seen in a desert as a black one. More of the black ones have been killed by lions and tigers, more of the dun colored have survived. The young deer grow like their parents in color. Deer are the same color on both sides because the deer of different colors are quickest seen and most often killed. Fewer of them live to have young.

Holden opposed this description of deer to the evolution of domesticated cows with their mottled coats whose coloring "makes no difference in their life on a farm." He also discussed lions and horses as examples of the processes of heredity, adaptation, and natural selection. Concerning horses, he wrote: "*The fittest survived* and had colts; *the weakest perished* in the struggle for existence" (emphasis in original).[38] Here the influence of Herbert Spencer's phrase "survival of the fittest" is evident in Holden's phraseology.

What is striking about these two books is how unusual they are among the corpus of texts that refer even tangentially to evolution. Of the several hundred science books for children examined for this project, only these two explained the concept of natural selection. Instead, many books for children focused on concrete descriptions of adaptations without situating those adaptations within the broader theory of natural selection or the concept of evolution.

A few other books used the word "evolution" without explaining natural selection, thereby avoiding engagement with controversies over whether Darwin, Lamarck, or other scientists' theories would prove predominant. One book was explicit in refusing to take a stand on whose theory was correct. Dana's *New Text-Book of Geology* claimed only that "the evolution of the system of life went forward through the derivation of species from species, according to natural methods not yet clearly understood."[39] In other books, the influence of evolutionary ideas was evident in their use of images and descriptions of nature that everywhere implied evolution without explaining the theory.

Evolution Implied

One example of a book that alludes to the scientific ideas behind evolution in various ways without explaining them is *The Earth in Past Ages* by Sophie Bledsoe Herrick, who was a science teacher in Baltimore before becoming an assistant editor for *Scribner's Monthly*. Her book was recommended in five different lists, in 1890, 1897, 1904, 1909, and 1915, second only to Darwin's own book. Herrick's book was classified in various lists under geology, science, out-of-doors, and stories of animals. In the text, Herrick asserted the importance of looking to geology to understand history, explaining: "Geology is the history of the earth and the various kinds of plants and races of animals that have lived upon it." She emphasized observation of natural forces such as steam, ice, and sedimentation, but then explained that studying the present allowed her readers to "look back to the past, and try to interpret the unknown by the known."[40]

Evolutionary theory and religion were not opposed in a simplistic way during this period, and those who made such polarizing arguments were typically "active participants in the conflict who wished to conceal the existence of a middle ground."[41] However, if the theory of evolution was right, there were real tensions over God's role in the universe. Darwin's theory of the role of random mutation in natural selection potentially made God essentially unnecessary to understanding the development of diverse forms of life. One of the religious counterarguments to the theory of evolution was that fossil evidence buried in the earth was not evidence of life in "past ages" but instead a divine test of human faith in God's role as creator. Herrick dismissed this: "But the world was not to be hoodwinked forever by such foolish superstitions." She refrained from greater theoretical explanation, aside from dismissing this religious explanation, but the implication is that she espoused evolution.

Herrick focused on familiar items such as salt, coal, and oil to anchor her exploration of geology in concrete terms. She included chapters on various animals, organized recognizably from lowest to highest in order, and concluded with a brief mention of humankind, albeit without explicit reference to evolutionary relationships between animals and humans. However, adaptation is implied in various instances, for example, "All life depends much on its surroundings; if they remain the same, the forms of life usually do not change much." She also strongly suggested evolutionary relationships between groups of creatures, as when she wrote that birds and reptiles were of one "family" but had been "separated so many millions of years that the family likeness is not very strong." In fact, Herrick even references Darwin's specific scientific contributions to such areas as understanding coral reefs and the discovery that "white cats with blue eyes are sure to be deaf."[42] Without explaining evolution, Herrick's book gave much information that prepared young readers for understanding the theory when they encountered it.

Evolution Suggested

In Darwin's *What Mr. Darwin Saw in His Voyage Round the World in the Ship "Beagle,"* evolution is implied in a brief discussion of mutual "adaptation in structure" between "a crab and a cocoa-nut tree."[43] This kind of allusion to the observable connections between creatures in the wild, either as consumer and food source or in more mutually beneficial relationships, is the most common way evolution appears in books for children from this period. While crabs and coconut trees were far afield from most children in the United States, there were observable connections between creatures more readily at hand, and those relationships were frequently used to illustrate adaptation.

Relationships between birds or insects and flowers appeared in several texts for children as imagery suggesting evolutionary adaptation. For example, Olive Thorne Miller's *Little Folks in Feathers and Fur* described "the bill of the hummingbird," as "a very curious thing. In each variety, it is shaped to suit the flower on which it feeds."[44] Similarly, Margaret Warner Morley's *A Song of Life* briefly mentioned evolution in a section that devoted extensive time to describing the reproductive structures of flowers. Morley anthropomorphized the flowers: "So keenly is this cross-fertilization desired by the flowers that they have evolved many curious devices to bring it about." She then proceeded to discuss how the structures of flowers attract bees as their needed pollinators.[45] Similarly, another of Buckley's books, *The Fairy-Land of Science*, did not explain evolution but instead hinted toward more complex mechanisms: "We cannot stop to inquire to-day how this all gradually came about, and how the flowers gradually put on gay colours and curious shapes to tempt the insects to visit them." Unlike Buckley's *A Short History of Natural Science*, this book treated evolution only indirectly, through references to the powers of nature that "adapted the flower to the insect and the insect to the flower."[46]

Other books suggested evolution by using adaptation as a concept and detailing the struggles of particular species to survive through climatic changes. In *Real Things in Nature*, Edward Holden's discussion of how acorns become oak trees includes adaptation. The author writes that even trees must struggle for survival against changing climates: "If the climate changes, the weak trees die and the others well adapt themselves to new circumstances. . . . Every living thing must adapt itself to its surroundings or die."[47] Similarly, Trotter suggests the evolution of camels is related to climatic and geographic changes: "The descendants of the camel-like beast that penetrated farthest in the Asiatic continent became, in the long course of time, adapted to a desert life, appearing in the present age as the camels of Bactria, Arabia, and North Africa." This was one of several examples in Trotter's *Lessons in the New Geography* of those species that "have greater facilities for overcoming barriers than others, some by their ready adaptation to changed conditions, their more varied diet and powers of resistance, others by superior means of locomotion, as with birds and many mammals."[48]

That the concept of evolution was repeatedly connected to imagery of plants and animals demonstrates one of the strategies science writers used to introduce complex ideas to children. This both reflected and contributed to cultural impact of nature study, with its stress on children's direct observations of the natural world. Authors emphasized observable relationships between creatures, anthropomorphizing those relationships in cases where they were mutually beneficial or focusing on struggle for survival in those cases where one species preyed upon another. These writers introduced highly contextualized versions of evolution and adaptation, providing a basis for children to understand more abstract versions of these concepts later in their education.

Human Evolution

Human evolution was the most controversial aspect of the theory of evolution, as it posed challenges to religious understandings of the divinely created and guided universe. In fact, several historians have argued that evolution controversies escalated precisely at moments when the theory was applied to humankind.[49] Since the 1925 Scopes trial over a textbook that featured evolution, debates over whether and how children should be taught about evolution have persisted, waxing and waning in intensity. Of those books that described or invoked some elements of evolutionary theory, by far the majority focused on the evolution of "lower" animals and left humankind out of the evolutionary picture. However, a few titles did explicitly situate humans in an evolutionary framework.

Frederic Arnold Kummer's book *The First Days of Man* was less explicit about evolution but included multiple references to adaptation. Kummer wrote that "no matter what sort of a life any creature is in the habit of living, if you make him live another kind of life, he will change himself to suit it." The third chapter, titled "The Ape That Walked Like a Man," described human evolution obliquely, by personifying Mother Nature, Earth, Sun, Wind, and other natural forces. These characters then discussed among themselves how to move the apes out of the forests and jungles and "up into the hills, where things will be different."[50] Personification of natural forces was a common trope in popular works of science for children.[51] In this case, it may have been a device to make the abstract concept of evolution more comprehensible to young readers.

Margaret Warner Morley's *A Song of Life* connected humans to the larger animal world by making explicit the developmental resemblances between human and fish embryos: "The human being, too, begins life as a single cell. He too passes through stage after stage of animal life, owning a far-away relationship to the simple creatures he so far outstrips. Gill openings convict him, too, of kinship with the fishes, and he passes through a stage, where, from one point of view, he looks absurdly like the embryo of a fish." Morley goes on to imply that humans are related to other mammals as well, passing through stages of embryonic development in which they "cannot be distinguished from an embryonic pig or a dog."[52]

The word choice here, that humans are "convicted" of "kinship," suggests some of the unease regarding how to understand humankind's place in a world of "lower" animals. Morley's focus on embryonic development in her writing suggests an affinity with the idea Ernst Haeckel promoted in association with Lamarckian evolution, that ontogeny, or the development of an individual organism, recapitulates phylogeny, or the evolution of the species. Still, nowhere in her book did the word "evolution" appear, and Morley's focus remained firmly on the growth of individual organisms rather than on the larger span of evolutionary time.

One book deserves special analysis for its radical changes in regards to inclusion of evolution over a series of editions. Edward Clodd's *The Childhood of the World* appeared in 1873 as a book for adults, but 1876 saw a special edition published for use in schools. These two editions were followed by many more, and by 1893 the book was in its tenth edition. Little change in the text was made throughout that time. Librarians' 1883, 1890, and 1893 lists recommended the book but listed it under general history, religion, and biology. These varied classifications indicate one central theme of Clodd's text: interweaving science and religion in understanding how humankind came to be. In these editions, he emphasized that God created and located most animals in appropriate places but that humans were special and different from the rest: "Where God has placed the brute, he has given it the covering best fitted for the place in which it lives, and has supplied its proper food close at hand. But God has placed man here naked, and left him to seek for himself the food and clothing best suited to that part of the world in which he lives."[53] This way of describing the order of the world echoes the book of Genesis, wherein Adam and Eve were cast out of the garden to make their own way in the world. However, it is also like the view promoted by Louis Agassiz, the famous nineteenth-century naturalist and influential speaker. Agassiz saw himself as a scientist but believed in a divinely created world wherein fossils were placed as mysteries and humans were created, alongside all other creatures, by the hand of God. One scholar has noted that Agassiz's ideas were not theoretical in the sense that Darwin's were, generating new lines of scientific inquiry, but instead were "edifices perched on top of mountains of data."[54]

However, in a 1914 edition of Clodd's book, he radically revised his way of writing about the history of humanity. This edition incorporated an explicit focus on evolution, with very little mention of God. Clodd included the phrase "natural selection." Now, rather than focusing on single act of divine creation, he wrote that "by slow steps and through long ages, the simplest living things have given rise to millions on millions of different plants and animals" and stated that "all plants and animals are made up of myriads of cells formed of the same stuff." More importantly, he explained a version of human evolution to his young readership: "There is no doubt that whatever that man and these big apes, and also the monkeys, sprung from a common ancestor . . . man has not come from an ape, as some ignorant people think; each has descended from a common ancestor." Like many scientists of the time, Clodd argued that "the common ancestor of Man and Apes had its home probably in some part of Asia." He also noted that "there were several kinds of half human creatures, as well as of manlike apes" in the fossil record.[55] The 1917 list of recommended books included Clodd's revised 1914 edition. Incidentally, another book by Clodd, his *Story of Creation: A Plain Account of Evolution* was the only recommended book of all nineteen examined that contained the word evolution in the title, and it appeared in one list in 1893.[56]

One book published in the early 1920s showed signs of changing cultural norms around teaching evolution to children, both featuring human evolution and explaining natural selection. Adam Gowans Whyte's 1921 *The Wonder World We Live In* celebrated Darwin as "one of the wisest men who ever lived" and included his picture in the preface. A later section, titled "Our Ape-Man Grandfather," explained human evolution in terms of natural selection: "So the answer to the question, 'where did the First Man come from?' is that 'he came from an ape-man!' And he came by the Better Brain Road. Just as in the stone age nature chose the cleverest men and killed off the stupid ones, so in the earlier age of ape man, nature favored the clever ape men and helped them to get on in the world. Man came from 'ape-men' by 'natural selection.'" This was immediately followed by a section titled "Are You Angry?" in which Whyte acknowledged that some people "get angry when they are told that men came from animals like apes." However, he compared these people to the priests who "refused to look through Galileo's telescope in case he should see something that would make him change his mind!" He ridiculed shame over being related to apes in an interesting way, appealing to children's love of pets: "Why should we be ashamed to have animals for our forty-second cousins? People are not ashamed to love their dogs and horses, and to admire them for their faithfulness and devotion. Some people, indeed, love their dogs and horses more than they love their own cousins." Whyte argues explicitly and directly for reason over religion, pointing out that fossil remains of "missing links" proved that "ape men became men by getting slowly cleverer and better through thousands and thousands of years."[57] By highlighting their fondness for animal companions, he also provided children with an emotional way to understand their distant kinship to animals.

Evolution appeared in partial and limited ways in most children's books, but this investigation provides a new context for understanding Van Loon's Newbery-award–winning *The Story of Mankind.* If the changing editions of Clodd's book and the 1921 book by Whyte are indicators, then it appears evolution was becoming a more common topic for children in the early 1920s. Viewed beside them, Van Loon's book is, arguably, continuous with a tradition of popular natural history for child audiences. Van Loon described an early hominid as a "creature, half ape and half monkey but superior to both" that was "the most successful hunter and could make a living in every clime. This creature, though you may hardly believe it, was your first 'man-like' ancestor."[58] Evolution is only the first topic in the book, which goes on to attempt a sweeping history of human societies, from Egypt and Mesopotamia all the way to the United States.

Although Van Loon won the most famous prize in children's literature, he had not intended to write for children. He thought he was writing a book for adults, albeit with copious illustrations. The audience for subsequent editions changed dramatically after children's librarians claimed this book for children. Other adult books of the time did explain evolution, including H. G. Wells's 1922 *A Short History of the World.* Van Loon's intended audience complicates the question of how children were exposed to evolutionary ideas, but it also suggests one possible explanation for why these ideas appeared so infrequently in books for children.

So Few Books, So Little Evolution

As mentioned, only thirty-two books of more than two hundred recommended science books described or alluded to evolution. The relative paucity of this evidence raises a question. Why is it that even before the political controversies raised by the Scopes trial so few books written for children contain or explain evolution?

The larger cultural context of science publication suggests one plausible explanation. There was a glut of popular science works for adults, many of them explaining evolution at a reasonably accessible reading level.[59] With so many popular, easily understood works written for adults, separate books for children below the ages of ten to fourteen might have seemed unnecessary. The content of librarians' lists supports this explanation. Adult books typically appeared in only one list, indicating they were not among the most canonical of librarians' recommendations. However, there were recommended books that explained evolution, as highlighted in their titles. A few examples include Ernst Haeckel's *The Evolution of Man*, Thomas Huxley's *On the Origin of Species and Evidence as to Man's Place in Nature*, Herbert Spencer's *The Factors of Organic Evolution*, and Charles Darwin's book *The Descent of Man and Selection in Relation to Sex*.

The publication context of children's science books suggests a second explanation: the evolution of humankind was not readily observable. When compared to the rambles in the woods or how-to books of biological specimen collection of nature study, evolution was theoretical to an extreme, involving vast spans of time. Some authors may have avoided the topic simply because it was difficult to explain to children. Whatever the reason, many authors chose science topics with a scope that was not conducive to discussion or description of evolution. For example, Elizabeth Agassiz, wife of Louis Agassiz, wrote *A First Lesson in Natural History* about aquariums and the animals that live in them. The text described a walk along the shore and the various creatures Agassiz found and collected there, addressed as a letter to her two nieces.[60] This is one of many examples of books that introduce science with a primary focus on observing the world around the child. Darwin's book for children had such a focus. Another example is Mrs. Alfred (Margaret) Gatty's book *Parables of Nature*, which appeared in only three of the lists but was an unusually successful book, remaining in print nearly one hundred years after its 1855 publication. Others include Ernest Thompson Seton's *Wild Animals I Have Known*, Olive Thorne Miller's *Queer Pets at Marcy's*, and Mrs. A. E. Anderson-Maskell's *Four Feet, Wings, and Fins: Natural Science for Young People*.

Evolution is only comprehensible in the context of geological time, and when the vast spans of geological time were represented at all in books for children, they were tied to very specific forms of observable evidence such as sedimentary rocks and rivers. For example, Edward Holden explained the age of the earth by reference to the Niagara River. "Rivers bring soil from the land at the rate of about one foot deep of soil every 5,000 years. . . . 6,000 times 5,000 = 30,000,000 years. If the rivers of old time worked no faster than the rivers of our time then the Earth must be at least 30,000,000 years old."[61] In another case, Arabella Buckley wrote that "people had so long held the belief that our earth had only existed a few thousand years" that the discovery of fossils was an immense surprise to them.[62] Sophie Bledsoe Herrick made explicit that the state of geological science was changing: "Geology is not an old science. It is scarcely one hundred years old to day. And some things are still unsettled and others unknown, but there are many things which are perfectly fixed and known and about the rest we are learning every day."[63] Clearly, these authors were not promoting a literal, biblical reading of the age of the earth, but neither were they offering a particularly accessible explanation of geological ages as a basis for understanding evolution.

Of course, there is the possibility that the controversial religious implications of evolution made it a touchy topic for children, and authors shied away. Despite this, a minority of authors did introduce or allude to evolution, so this avoidance must be read as a tendency rather than a taboo. Historians Ronald Numbers and Lester Stephens point out that controversy over evolution escalated when the theory was specifically applied to humans.[64] The books by Clodd and Whyte by themselves cannot be taken to indicate a publishing trend, but they are definite instances of teaching human evolution to children. Further research might yield a fruitful picture of how book recommendations changed in the years after the 1922 Newbery medal, during a period of increased controversy.

Conclusion

The political and cultural landscape of the United States has changed dramatically since the early twentieth century, but controversies persist around the topic of teaching evolution to children. What emerges from this examination is evidence that evolution was present in the whole corpus of science recommended for children, albeit in limited numbers of books recommended specifically to a child readership. This investigation raises larger questions about how adults influenced children's access to the concept of evolution. What stories were children presumed to be able to understand, and what was considered beyond their comprehension? When were children assumed ready to learn the more complex and theoretical stories of science? Examination of other sources, including records of science teachers from the period, might yield complimentary evidence to construct a more complete historical picture.

In the history of children's reading experiences, evolution is only one of many controversial topics. How it is treated in children's literature provides one answer to the larger question: how and when are children granted access to controversial topics? The method employed for this analysis, using librarians' lists to identify relevant texts, is only one means of addressing this question. Librarians' lists have a distinct advantage: they approximate what children could have actually read. They don't constitute evidence of reading, but they are one form of evidence that certain books were available to children, at least in public libraries. Examination of actual library collections and circulation records would provide a better sense of the extent of their influence.

These lists have the advantage of representing the opinions of relatively centralized professional authority. They were affiliated with the ALA, youth services leader Caroline Hewins, or recognized library publisher H. W. Wilson, and they appeared at a time librarians across the nation were asking themselves and each other, at conferences and even in surveys of their colleagues, how to provide services to children and what those services should be.[65] There is much room for other approaches that lead to the discovery of historical evidence of how real children encountered evolution and other controversial topics in texts. Further research might determine how these lists related to aspects of the wider market for children's science books such as advertising for these books or numbers sold.

Children's science trade books conveyed evolution only occasionally and in limited ways. While there were controversies around evolution and religion, close examination of the thirty-two texts that mentioned evolution in some form suggest practical rather than political explanations for what was, and was not, included. When children read about the natural world and evolution, they read about insects, flowers, animals' special adapted appendages, and other observable natural facts. Few authors attempted to convey to their child readers the sweeping scale of geological time that was fundamental to developing a broad understanding of evolution or natural selection. Instead, most who wrote for children focused on affinities between their young readers and the natural world, where animals became pets and evidence of adaptation was part of a world of wonder, where evolution was everywhere implied but rarely explained.

Appendix: The Thirty-Two Books Appearing in the Ten Lists That Included Reference to Darwin or Evolution

Bert, Paul. *First Steps in Scientific Knowledge*. Philadelphia: J. B. Lippincott Co., 1886.
Buckley, Arabella B. *Fairy-Land of Science*. New York: D. Appleton and Co., 1905.
——. *Life and Her Children*. London: Edward Stanford, 1881.
——. *A Short History of Natural Science*. New York: D. Appleton and Co., 1876.
——. *Through Magic Glasses and Other Lectures*. New York: D. Appleton and Co., 1881.
——. *Winners in Life's Race; or, The Great Backboned Family*. New York: D. Appleton and Co., 1893.

Champlin, John D. *The Young Folks' Cyclopædia of Natural History*. New York: Holt and Co., 1905.

Clodd, Edward. *The Childhood of the World: A Simple Account of Man in Early Times*. London: Kegan Paul, Trench, Trubner & Co., Ltd., 1893.

———. *The Childhood of the World: A Simple Account of Man's Origin and Early History*. New Edition, Rewritten and Enlarged. New York: Macmillan Co., 1914.

———. *A Primer of Evolution*. New York: Longmans, Green, and Co., 1895.

Dana, James D. *New Text-Book of Geology*. New York: Ivison, Blakeman, Taylor, 1874.

Darwin, Charles. *What Mr. Darwin Saw in His Voyage Round the World in the Ship "Beagle."* New York: Harper & Brothers, 1879.

Dopp, Katharine E. *The Tree Dwellers*. Chicago: Rand McNally and Co., 1903.

Grey, Elisha. *World Building and Life: Earth, Air and Water*. Vol. 1. New York: Fords, Howard, and Hulbert, 1899.

Happold, F. C. *The Adventure of Man*. London: Christophers Ltd., 1926.

Herrick, Sophie B. *The Earth in Past Ages*. New York: American Book Co., 1888.

Hillyer, V. M. *A Child's History of the World*. New York: Appleton-Century-Crofts, Inc., 1924.

Holden, Edward Singleton. *Real Things in Nature*. New York: Macmillan, 1903.

Hornaday, William T. *American Natural History*. New York: Charles Scribner's Sons, 1904.

Hutchinson, H. N. *The Autobiography of the Earth*. New York: D. Appleton and Co., 1891.

Johonnot, J. *Neighbors with Claws and Hoofs*. Natural History series, number 3. New York: D. Appleton and Co., 1885.

Kummer, Frederic Arnold. *The First Days of Man*. New York: George H. Doran Co., 1922.

Miller, Olive Thorne. *Little Folks in Feathers and Fur*. New York: E. P. Dutton & Co., Inc., 1880.

Minot, Henry Davis. *Land and Game Birds of New England*. New York: Houghton, Mifflin, and Co., 1903.

Morley, Margaret Warner. *A Song of Life*. Chicago: A. C. McClurg & Co., 1909.

Trotter, Spencer. *Lessons in the New Geography*. Boston: D. C. Heath & Co., 1895.

Van Loon, Hendrik W. *Ancient Man: The Beginning of Civilization*. New York: Boni and Liveright, 1922.

———. *Story of Mankind*. New York: Boni and Liveright, 1921.

Wells, Margaret E. *How the Present Came from the Past*. New York: Macmillan, 1917.

Whyte, Adam Gowans. *The Wonder World We Live In*. New York: Alfred A. Knopf, 1921.

Wood, J. G. *Illustrated Natural History*. London: Routledge, Warne, and Routledge, 1862.

Wright, Henrietta Christian. *Children's Stories of the Great Scientists*. New York: C. Scribner's Sons, 1888.

Notes

I gratefully acknowledge the financial assistance of the Graduate School of Library and Information Science (GSLIS) and the research assistance of graduate students Anna Dombrowski, Minjie Chen, and Alaine Martaus. I also thank my colleagues for feedback from the Print Culture and STEM conference audience and the GSLIS Research Writing Group, led by Dr. David Dubin.

1. Irene Smith, *A History of the Newbery and Caldecott Medals* (New York: Viking, 1957), 140.

2. Hendrik Willem Van Loon, *The Story of Mankind* (Chapel Hill, N.C.: Yesterday's Classics, 2007), 16.

3. Sidney I. Dobrin and Kenneth B. Kidd, *Wild Things: Children's Culture and Ecocriticism* (Detroit: Wayne State University Press, 2004), 308.

4. Sally Gregory Kohlstedt, "Nature, Not Books: Scientists and the Origins of the Nature-Study Movement in the 1890s," *Isis* 96, no. 3 (2005): 324–52.

5. Ronald P. Ladouceur, "Ella Thea Smith and the Lost History of American High School Biology Textbooks," *Journal of the History of Biology* 41, no. 3 (2008): 435–71.

6. Ronald L. Numbers and John Stenhouse, *Disseminating Darwinism: The Role of Race, Place, Religion, and Gender* (New York: Cambridge University Press, 1999), 300.

7. In 1900, the first Children's Librarians' Section of ALA was formed at the same time that the first full training program for children's librarians opened at the Pittsburgh Carnegie Public Library, marking the formalization of this specialty within librarianship.

8. The books examined here were all produced and collected in the United States. Recent research indicates that in some cases books imported from the British to the American context were changed very extensively. For more on comparisons of fiction, see Leslee Thorne-Murphy, "Re-Authorship: Authoring, Editing, and Coauthoring the Transatlantic Publications of *Charlotte M. Yonge's Aunt Charlotte's Stories of Bible History*," *Book History* 13 (2010): 80–103; Gillian Avery, *Behold the Child: American Children and Their Books, 1621–1922* (Baltimore: Johns Hopkins University Press, 1994), 226.

9. Caroline M. Hewins, *Books for the Young: A Guide for Parents and Children* (New York: Leypoldt, 1882).

10. For more information about Hewins's authority as leader in early children's librarianship, see Kate McDowell, "The Cultural Origins of Youth Services Librarianship, 1876–1900" (PhD diss., University of Illinois, 2007); Anne H. Lundin, "A Delicate Balance: Collection Development and Women's History," *Collection Building* 14, no. 2 (1995): 42–46; Sybille A. Jagusch, "First among Equals, Caroline M. Hewins and Anne C. Moore: Foundations of Library Work with Children" (PhD diss., University of Maryland, 1990).

11. Kate McDowell, "Children's Voices in Librarians' Words: 1890–1930," *Libraries and the Cultural Record* 46, no. 1 (2011): 73.

12. Children's science books first received specialized reviewer attention in the 1925 edition of *Children's Catalog*, in which "specialists" in "technical lines such as chemistry, electricity, radio, etc." added their approval to librarians' selections of best books. See Minnie Earl Sears and Corinne Bacon, eds., *Children's Catalog: A Dictionary Catalog of 4100 Books with Analytical Entries for 863 Books*, 3rd ed. (New York: H. W. Wilson, 1925); Hewins, *Books for the Young* (1882), 82; Caroline M. Hewins, *Books for Boys and Girls: A Selected List* (Boston: Library Bureau, 1897), 31; Caroline M. Hewins, *Books for Boys and Girls: A Selected List*, 2nd rev. ed. (Boston: American Library Association Publishing Board, 1904), 56; Caroline M. Hewins, *Books for Boys and Girls: A Selected List*, 3rd rev. ed. (Chicago: American Library Association Publishing Board, 1915), 5.

13. Kate McDowell, "Which Truth, What Fiction? Librarians' Book Recommendations for Children, 1876–1890," in *Education and Print Culture in Modern America*, ed. Adam Nelson and John Rudolph (Madison: University of Wisconsin Press, 2010), 15–35.

14. Hannah P. James, "Yearly Reports on the Reading of the Young," *Library Journal* 10, no. 8 (1885): 291; Mary A. Bean, "Report on Reading of the Young," *Library Journal* 8, nos. 9–10 (1883): 220.

15. John Frederick Sargent, Mary E. Sargent, and Abby L. Sargent, *Reading for the Young: A Classified and Annotated Catalog, with an Alphabetical Author-Index* (Boston: Published for the American Library Association Publishing Section by the Library Bureau, 1896), 225; John Frederick Sargent, *Supplement to Reading for the Young: A Classified and Annotated Catalogue with an Alphabetical Author-Index* (Boston: Library Bureau, 1890), 3.

16. American Library Association, *Catalog of "A.L.A." Library: 5000 Volumes for a Popular Library* (Washington, D.C.: GPO, 1893); Corinne Bacon, *Children's Catalog of Thirty-Five Hundred Books: A Guide to the Best Reading for Boys and Girls* (White Plains, N.Y.: H. W. Wilson Co., 1917); *Children's Catalog* (White Plains, N.Y.: H. W. Wilson Co., 1909).

17. Hewins, *Books for Boys and Girls* (1897), 7; Hewins, *Books for Boys and Girls* (1914), 48.

18. A full list of search terms is "Darwin," "evolution," "evolutionary," "selection," "natural selection," "adaptation," adapted," "struggle," "struggled," "survive," "survived," "survival," "fit," "fittest."

19. Frank Buckland, *Log-Book of a Fisherman and Zoologist* (London: Chapman & Hall, 1891), 338.

20. Dominick Cavallo, *Muscles and Morals: Organized Playgrounds and Urban Reform, 1880–1920* (Philadelphia: University of Pennsylvania Press, 1981), xiv, 188.

21. See Sally Gregory Kohlstedt, "'Through Books to Nature': Texts and Objects in Nature Study Curricula," this volume.

22. Louisa May Alcott, *Little Men: Life at Plumfield with Jo's Boys* (Boston: Little, Brown, 1899), 181.

23. McDowell, "Which Truth, What Fiction?" 23.

24. Jay Martin, *The Education of John Dewey: A Biography* (New York: Columbia University Press, 2002), 562.

25. Kohlstedt, "Nature, Not Books."

26. Jean-Marc Drouin and Bernadette Bensaude-Vincent, "Nature for the People," in *Cultures of Natural History*, ed. Nicholas Jardine, James A. Secord, and Emma C. Spary (Cambridge: Cambridge University Press, 1996), 410; Bernard V. Lightman, *Victorian Science in Context* (Chicago: University of Chicago Press, 1997), 188.

27. Drouin and Bensaude-Vincent, "Nature for the People," 408–25.

28. Lightman, *Victorian Science in Context*, 3–6.

29. Barbara T. Gates, "Revisioning Darwin with Sympathy: Arabella Buckley," in *Natural Eloquence: Women Reinscribe Science*, ed. Barbara T. Gates and Ann B. Shteir (Madison: University of Wisconsin Press, 1997), 164–76.

30. Howard P. Chudacoff, *How Old Are You? Age Consciousness in American Culture* (Princeton, N.J.: Princeton University Press, 1989), 232; Karen Sanchez-Eppler, *Dependent States: The Child's Part in Nineteenth-Century American Culture* (Chicago: University of Chicago Press, 2005), 260; Joseph F. Kett, *Rites of Passage: Adolescence in America, 1790 to the Present* (New York: Basic, 1977), 327.

31. Edward S. Morse, *First Book of Zoology* (London: Henry S. King & Co., 1876), 190.

32. Henrietta Christian Wright, *Children's Stories of the Great Scientists* (New York: C. Scribner's Sons, 1888), 350.

33. Arabella B. Buckley, *Life and Her Children* (London: Edward Stanford, 1881), 176.

34. Arabella B. Buckley, *A Short History of Natural Science* (New York: D. Appleton and Co., 1876), 467.

35. Spencer Trotter, *Lessons in the New Geography*, 2nd rev. ed. (Boston: D. C. Heath & Co., 1895), 4, 163, 165.

36. Charles Darwin, *What Mr. Darwin Saw in His Voyage Round the World in the Ship "Beagle"* (New York: Harper & Brothers, 1879), 10, 17, 19.

37. Buckley, *Short History of Natural Science*, 465, 474–75.

38. Edward Singleton Holden, *Real Things in Nature* (New York: Macmillan Co., 1910), 175.

39. James Dwight Dana, *New Text-Book of Geology: Designed for Schools and Academies*, 3rd ed., rev. and rearranged (New York: Ivison, Blakeman, Taylor, 1874), 347.

40. Sophie Bledsoe Herrick, *The Earth in Past Ages* (New York: American Book Co., 1888), 1, 7.

41. Peter J. Bowler, *Evolution: The History of an Idea*, 3rd ed., completely rev. and expanded (Berkeley: University of California Press, 2003), 203.

42. Herrick, *Earth in Past Ages*, 5, 203, 209–10, 94, 133.

43. Darwin, *What Mr. Darwin Saw*, 89.

44. Olive Thorne Miller, *Little Folks in Feathers and Fur* (New York: E. P. Dutton & Co., Inc., 1880), 141.

45. Margaret Warner Morley, *A Song of Life*, 9th ed. (Chicago: A. C. McClurg & Co., 1909), 43, 37.

46. Arabella B. Buckley, *Fairy-Land of Science* (New York: D. Appleton and Co., 1905), 215, 208.

47. Holden, *Real Things in Nature*, 176–77.

48. Spencer Trotter, *Lessons in the New Geography*, 2nd rev. ed. (Boston: D. C. Heath & Co., 1895), 6–7.

49. Numbers and Stenhouse, *Disseminating Darwinism*, 129.

50. Frederic Arnold Kummer, *The First Days of Man* (New York: George H. Doran Co., 1922), 32, 43.

51. Lightman, *Victorian Science in Context*, 189.

52. Morley, *Song of Life*, 141–43.

53. Edward Clodd, *The Childhood of the World: A Simple Account of Man in Early Times*, 10th ed. (London: Kegan Paul, Trench, Trubner & Co., Ltd, 1893), 10.

54. Louis Menand, *The Metaphysical Club* (New York: Farrar, Straus and Giroux, 2001), 141.

55. Edward Clodd, *The Childhood of the World: A Simple Account of Man's Origin and Early History*, new ed., rewritten and enlarged (New York: Macmillan Co., 1914), 11, 13, 16, 21.

56. Edward Clodd, *Story of Creation: A Plain Account of Evolution* (London: Longmans, Green, and Co., 1888), 242.

57. Adam Gowans Whyte, *The Wonder World We Live In* (New York: Alfred A. Knopf, 1921), 4, 121–24.

58. Van Loon, *Story of Mankind*, 4–8.

59. Drouin and Bensaude-Vincent, *Nature for the People*, 408–25; Lightman, *Victorian Science in Context*, 489.

60. Elizabeth Agassiz, *A First Lesson in Natural History*, 2nd ed. (Boston: Little, Brown and Co., 1859), 82.

61. Holden, *Real Things in Nature*, 140–41, 165.

62. Buckley, *Short History of Natural Science*, 404.

63. Herrick, *Earth in Past Ages*, 6.

64. Numbers and Stenhouse, *Disseminating Darwinism*, 300.

65. Kate McDowell, "Surveying the Field: Women's Research for Youth Services in Professional Librarianship, 1882–1898," *Library Quarterly* 79, no. 3 (2009): 279–300.

"Through Books to Nature"

Texts and Objects in Nature Study Curricula

SALLY GREGORY KOHLSTEDT

Educators in the early twentieth century faced the dilemma of how to build the skills of teachers so that they could teach directly from nature in a new progressive pedagogy emerging in the late nineteenth century known as nature study. How should printed materials be used when the goal was to advance the hands-on study of natural specimens and nature in situ? The paradox of emphasizing learning about nature was best accomplished by working with objects while at the same time producing texts describing nature purporting to encourage observational practice was not lost on nature study advocates. Those who wrote guides for educators moved carefully through this problem as they sought to explain and demonstrate the new curriculum. While it seemed feasible, perhaps, to follow Louis Agassiz's injunction to "study nature, not books" when thinking about classroom pupils or even college students at Harvard, how were schoolteachers to be prepared to guide them?[1] Textbook authors Charles B. Scott, Clifton F. Hodge, and Anna Botsford Comstock approached the subject in quite distinct ways, but their publications offered schoolteachers techniques and examples intended to stimulate pedagogical imagination and avoid simple emulation. Their books and materials were among the most widely circulated when nature study flourished from about 1895 to 1920, frequently assigned in normal school classes across the country, circulated in teachers' reading circles, and used by individual teachers as they prepared their classes.

Progressive educators—regardless of their attention to urban or rural schools, concern with primary or grammar grades, or geographic locations rich in varied wildlife or highly dense in human population—generally agreed that the best way to teach was to rely on pupils' tactile encounters with nature. The emphasis on using actual specimens had grown out of "object teaching" in the 1870s and 1880s but took a particular turn away from what many came to view as a static technique that relied on teacher-prepared materials.[2] By contrast, the new curriculum of nature study insisted on starting with children's curiosity and relied on observation by individual pupils, often exploring in groups and under guidance from a teacher. In this context, natural specimens were often studied in their local environment and were purposefully made part of preparation for future studies. But this dramatic shift in curriculum toward hands-on nature study in the public schools raised multiple questions: What was the role of the teacher? Should textbooks and illustrations be used? If so, when and how should they be introduced? Basically, where did print fit into nature study activities?

Publishers were quick to respond to the so-called new education curriculum, including nature study, at the end of the nineteenth century. Emerging textbook companies and more general publishers competed fiercely for the business being generated by the growing number of normal (or teachers' training) schools, which, in turn, reflected increasing school enrollments across North America.[3] Most of the textbook authors were mainstream, white, and primarily northern educators. Many were instructors in teacher training or normal schools, but a few were practicing teachers eager to share their own experiences even as they generated materials for the new nature study curriculum. Some wrote because publishing could be lucrative, and others sought a platform from which to advocate for a philosophical outlook as well as to share their own effective methods and content. All were sure teacher education would require books for transmitting basic information about the natural world and demonstrating potential techniques for investigating it. Textbooks were, after all vetted information presumed reliable and able to provide core information useful to multiple users. Textbook providers could not, however, presume uniform reception.[4]

The argument that young pupils should study nature directly stemmed from several sources in the 1890s. Particularly evident at the outset were a number of scientists who represented disciplines that had grown increasingly strong and influential. Louis Agassiz, a popular lecturer, teacher, and museum advocate at mid-century, had famously inspired his own students at Harvard by having them spend their first hours in his lab closely analyzing a single fish specimen and had carried his techniques to the seaside in his famous school for educators at Penikese Island in the 1870s.[5] Other prominent educators, like Harvard's outspoken president, Charles Eliot, campaigned for more precollege science preparation in academies and high schools as he challenged classic subjects and encouraged electives at the university level. A scientist by training, Eliot was readily persuaded that introducing even very young children directly to nature was important. Many scientists became early supporters of the movement, and a few contributed textbooks.[6] Such academic educators influenced the generation that followed them and took seriously the work of even elementary teachers.

While scientists pushed this agenda on the lecture circuit, in popular magazines, and at professional meetings, a second group of academic educators approached the issue with a very different set of concerns, namely changing ideas about pedagogy and the nature of early education. At the substantial normal schools built across the nation, training for teachers in the 1880s and 1890s became increasingly sophisticated theoretically and more explicit about applications of this theory.[7] Normal school instructors introduced research on educational psychology, much of it written by new PhDs trained in Europe or available through translated essays and books that emphasized the importance of child development. Some theorists were influenced by recapitulation theory (i.e., individual development echoed that of the race or culture) and argued that young children were, like ancient peoples, especially close to nature. Other social scientists brought in quantitative assessments of learning capacity. Psychologist G. Stanley Hall, best remembered for his introduction of the concept of adolescence, did his earliest work with even younger children.[8] Using testing techniques borrowed from William Wundt, he analyzed how children learned and what they retained; he stressed that such information would help teachers attend to the specific interests and capabilities of children by age cohort. He and others in the child study movement argued that it was important to study learning in a scientific way. Influential educational experts like Hall at Clark University, John Dewey at the University of Chicago, and Edward Thorndike at Teachers College, Columbia University, proffered competing theories about child development, but they all agreed that older styles of role learning and memorization of textbook materials were ineffective; young pupils were not miniature adults.[9]

Nature study offered an opportunity to put into theory into practice with new educational methods, especially experiential learning, collaborative projects, and integrated curricula. Teaching with objects had been introduced in the 1860s in the United States, but by the 1890s, the method was critiqued as artificial, narrow, and didactic. Nonetheless, some of its tactile qualities had proven effective in teaching, so using objects identified by pupils as curious, interesting, or significant was a reasonable extension of the older approach. But nature study posed a different challenge. If most teachers had not been taught about science and the natural world, how could they be expected to teach such subjects? And wasn't it a contradiction to use books to teach about the nature that children were to learn by observing and handling natural objects?

Books had always been a standard teaching tool and remained important. According to historian of education Carl Kaestle, in the 1890s America experienced a "second revolution" in publishing. Literacy in the United States was high, with a self-reported illiteracy at just 6.2 percent in 1900.[10] Newspaper, magazine, and book publishing flourished in a highly competitive market even as public libraries made these materials available to presumably even poor populations in urban areas through outreach programs and the introduction of special children's rooms with open bookshelves.[11] Major and small publishers produced textbooks, teachers' manuals, student readers, pamphlets, and other educational material, leading one normal school instructor (and textbook writer) to observe, "The schools have been flooded with Nature Primers and Nature Readers, and Hours with Nature, good, bad, and indifferent to meet the demands of the school-men and school-women."[12] But where did printed materials fit into nature curriculum—and, indeed, what nature should pupils study?

The term "nature study" had caught on quickly after Wilbur Jackman used it at Cook County Normal School and then produced a detailed textbook for his teachers in training that outlined what he called a "rolling year" of curricular projects that followed the seasons. Trained at a normal school in Pennsylvania and with a bachelor of science from Harvard in 1884, Jackman produced a broad and scientific schema—nature encompassed the physical world through the human experience.[13] His ideas circulated quickly in educational periodicals and meetings of the National Education Association, as the subject gained currency among other new subjects, particularly geography and history. Equally important, he produced a textbook directed at teachers that explained the nature of the curriculum and emphasized the method of teaching using objects, observation, and direct tactile experience. His *Nature-Study for the Common Schools* was directed to the "common school teacher" who would implement such work in the classroom.[14] He was aware that reading was inevitable and important for pupils, but observation should come first. So, Jackman encouraged his pupils at the practice school of Cook County Normal School to describe their observations, write them on the blackboard, and read them aloud. At the end of the day, the work would be transcribed to paper and sent to the printing office to become a reading lesson for all. As Jackman put it, the child thus learned that reading was "a natural supplement to his own thought and observation."[15] This provided, as well, awareness that reading gave access to someone else's experience but in itself did not constitute final authority.

Having learned his science from books and direct instruction, Jackman presumed teachers would learn the same way, in addition to whatever experience they might have in a practice or experimental school associated with a normal school. Teachers would use their acquired competence to affirm accurate observations of the pupils. Thus, they would overcome what he believed was the most serious obstacle facing individuals who were assigned or sought to teach about the natural world, namely their lack of faith in their own "ability to do anything useful or creditable."[16] His method required that children in their earliest school years, at least, build on their own curiosity and explore using all their senses. While Jackman's overview, and indeed his month-by-month recommendations for course content, were quite directive in his recommendations to teachers about subjects and approaches, nature study quickly showed its versatility because it relied on teachers using materials in their local environment. Other books were soon in competition with Jackman's as the subject gained national attention.

Textbooks used by teachers as well as those addressed to pupils deserve systematic attention from historians because they reveal the ideology surrounding a particular subject and identify perceived best practices.[17] Very little attention has been directed at the topic generally, although scholars specifically interested in children's literature have discussed the widespread interest of books about animals and plants, noting a growing enthusiasm for nature stories and handbooks about birds, minerals, and other subjects in the last half of the nineteenth century. Much of the nature study material for children was in the form of leaflets that reflected local or regional natural life and topography, as when G. L. Connor Jr. of Denver produced a series that ranged from "Denver Soils" to "Mountain Lakes.[18] Some of the best and most popular stories about nature for children were, as Barbara Gates notes, realistic even as they pushed toward illusion and metaphor that would "expand limited human sensory perception."[19] This enthusiasm paralleled that of adults at the turn of the century, when environmentally conscious authors like John Muir, John Burroughs, and Ernest Seton Thompson were among the best sellers and often used in teachers' reading circles.[20]

Nature-study textbooks, however, represented a puzzle. As primarily advice books for teachers, they were different from those for reading, arithmetic, and other subjects presented in a relatively prescriptive way. Nature study authors needed to negotiate the issue of preparing teachers adequately to teach about nature while trying to avoid establishing a narrowly didactic course that would mitigate against the very methods of investigation and use of objects that were at the core of its philosophy. Edward Bigelow, a lecturer on nature study, tried to describe the tension. He turned to poets like William Wordsworth, who wrote eloquently about his emotional response to nature and warned that written words could dull the senses. Bigelow agreed, yet sought to find ways to reconcile the potential antagonism yet dynamic sensibility between books and nature. He found some resolution in an author who concluded, "Books and nature are never inimical; they mutually speak for and interpret each other; and only he who stands where their double light falls sees things in true perspective and in right relations."[21] That abstract conceptualization of a "double light" did not, however, offer a ready solution to teachers who sought advice about both what to teach and how. Nature study authors also wrestled with a parallel problem, that of introducing expert knowledge.

Most authors of widely distributed textbooks brought specialized knowledge to their projects, with bachelor's and sometimes doctoral degrees in science. Many also established credibility through their personal experience teaching in precollegiate classrooms. They envisioned themselves as purveyors or translators of the nature study idea even as they explained their pedagogical techniques and offered quite practical advice about teaching young pupils. Their books were among those handbooks most often put in place and tested in the laboratory/experimental/practice schools affiliated with normal schools in the early twentieth century. These were not the only textbooks, and administrators in large urban systems produced detailed courses of study, written to ensure that graded classes proceeded in parallel with one another. In some cities, nature study supervisors were appointed to help identify and provide appropriate specimens, teach special classes for teachers, and even lead weekend excursions to potential sites for classroom visits. All the textbook authors encouraged teachers to provide pupils with natural objects and to take nature walks.

Among the dozen or more widely used textbooks, those by Charles Scott, Clifton Hodge, and Anna Botsford Comstock were published within about a decade of each other and at the moment when the trajectory of nature study courses was rising steeply, just at the turn of the century. Having taught school, they wrote in a way that presumed that teachers, their primary audience, were their peers and collaborators in advancing and implementing the nature study program. What seems to have made their books particularly useful was their balance between what were called "methods" for teaching and the content or topics recommended for implementing these. This combination was also prevalent in normal school curriculum, where a course on nature study that introduced fundamental botany, zoology, geology, and other topics could be paired with one on the pedagogy of nature study.[22]

Writing textbooks proved a convenient way to earn additional income and publicize a particular outlook on nature study. Charles Scott, served on a subcommittee of the National Education Association's prestigious Committee of Ten, charged to rework and standardize the high school curriculum.[23] It focused primarily on identifying required courses for four-year public high schools in a standard set that would be adopted nationwide. Its recommendations started a movement to make high schools sufficiently similar that students could transfer among them and colleges could presume a certain level of achievement among graduates. Scott was assigned to the natural history subcommittee, and within it he took responsibility for recommending pre-high school preparation that would initiate younger students to scientific subjects. His outline of a proposed curriculum for elementary and grammar schools, produced in 1893, was identified as nature study and echoed the topics Jackman had suggested in his rolling year curriculum.[24] Within three years, Scott had accepted a position at the New York State Normal School in Oswego. That move positioned him to formulate a substantial textbook for normal school students, which combined theory, method, and examples in its 618 pages. His presumed readership brought varied educational backgrounds, ranging from those who had only eighth-grade educations to those who had taught for a number of years and others who had attended an academy or college. He could and did assume that relatively few of them had been taught much science or, if they had, that their preparation entailed memorization of facts rather than an outlook on nature. His goal was to help prospective teachers learn the philosophy of nature study as well as potential content.

His textbook opened with an extended discussion of the dandelion, a highly available subject (figure 13). Familiar to most students as a "despised, wayside weed," Scott used it to reopen the issue of objects, this time stressing those found in nature. Students were each to pull a dandelion to study carefully. This common plant would allow them, as they examined its parts and compared them to other flowing plants, to learn the rudimentary components, including root, stem, and leaf shape as well as the stamen and pistil of the flower. But that discussion was only the beginning of thinking about the way the often dismissed plant actually invaded yards and self-seeded in unwanted places; dandelions offered an opportunity to consider the geography of plant distribution. In addition, teachers could ask about usage, encouraging pupils to describe how the leaves could be eaten as a healthy, inexpensive vegetable and mentioning that the plant had medicinal properties. If that range of related topics were not enough to reposition the dandelion in pupils' thinking, Scott pointed out that not all people had viewed the dandelion negatively, and he provided stanzas from poets like James Russell Lowell, who described the yellow flower's bright, engaging appearance. This, then, was a nature study project that could be done out-of-doors by even very young pupils. The dandelion was ubiquitous and provided an integrative topic readily related to geographic, social, and economic topics in the curriculum—a technique named the "correlation method" by contemporary educational psychologists who supported and extended the theories of Friedrich Herbart in the late nineteenth century.[25] Correlation went in two directions, and a number of authors produced readers that introduced vocabulary and reading skills while complementing nature study projects.[26]

The title of Scott's textbook was significant. *Nature Study and the Child* had a long section on educational psychology and child development, thus fitting well into a single normal school course intended to integrate pedagogical methods with content. Scott demonstrated that contemporary educational theories could be readily implemented through nature study by activities specific to the particular ways children learned at various ages and stages of development. By familiarizing prospective teachers with the theoretical insights and pedagogical suggestions of European and North American educational theorists, Scott intended to open up the prospective teachers' capacity to frame their nature

Figure 13. Anna Botsford Comstock made this
lovely wood engraving of dandelions and a
locust for her husband, John Henry Comstock's
Insect Life: An Introduction to Nature Study (1899),
plate 3, facing page 167.

study projects in ways appropriate for young children. He left little to chance,
however, in terms of fundamental knowledge, and the second half of the book
presented detailed content relating to plants, animals, minerals, to astronomical
and meteorological observations, and to principles of physics. For Scott, *Nature
Study and the Child* was to be a sourcebook on nature but one that simultaneously
provided examples of how a teacher might awaken children's literary, artistic,
and spiritual sensibilities.

Scott's tone was collaborative as he wrote, "It behooves us as teachers to
prepare ourselves"; but he switched to the voice of an expert in discussing, in

more abstract terms, the "mental development of the child."[27] His tone and style were apparently successful. Data on publication and distribution are difficult to find, but his book, published by D. C. Heath, went through at least four editions, was often on the recommended lists of books in normal schools, and was, of course, the standard at Oswego.[28] Wilbur Jackman's original 1891 nature study book had been directive in its program, lacked illustrations, and concentrated on content to be taught without much comment on pedagogy beyond pointing to specific topics that would work well throughout the calendar year. While he was typically recognized as a founder of the movement that emphasized teaching children directly in nature, his textbook seems to have quickly faded from active use as others, some from a new generation of authors, came on the market. These successors were closely attuned to changing educational theory and focused on quite different, often imaginative ways of bringing material objects into student learning.[29]

One textbook, well-reviewed in *Science* magazine, that particularly captured the attention of practicing teachers was Clifton Hodge's *Nature Study and Life*.[30] Hodge had taught school briefly but then taken a PhD in psychology at the young Johns Hopkins University. In 1892 he joined Hall at Clark University, where he helped edit *Pedagogical Seminary,* the first journal of educational psychology in the United States. Clark University, established with the ambition of being among the prominent private research universities, did not have an education department with a practice or experimental school like those developed at the University of Chicago and Teachers College, Columbia. Instead, Hodge collaborated with teachers in nearby public schools in Worcester, Massachusetts. Here, enthusiastic teachers enabled him to test some of his ideas about nature study and think in very practical ways about the struggle to teach in the difficult educational environment of crowded slums in an industrial city where pupils had few obvious resources. His matching of real-life experiences with textbook writing provided a distinctive demonstration that nature study could be effectively implemented in what originally seemed unlikely circumstances of crowded inner-city slums, far from the verdant spaces found in smaller towns or middle-class neighborhoods Scott envisioned.

Hodge had himself grown up in a rural environment and fully appreciated what his childhood had taught him about nature in woodlands and streams, but he was pragmatic about the industrial realities around him. He responded to the challenge by designing for the Worcester pupils quite different projects that took into account their social reality and a natural world fundamentally altered by human intervention. Like Jackman and Scott, Hodge sought to provide advice on teaching about plants and animals, but his approach was innovative as he demonstrated to pupils that such knowledge of nature around them was applicable in their daily lives, too.

Probably influenced by his colleague G. Stanley Hall, Hodge approached the subject very deliberately from a child-development perspective. He encouraged using the immediate environment that influenced young people's perceptions of nature and over which they had some influence. City children most often encountered living animals as pets, so he recommended starting nature study each year by talking about those of pupils and then encouraging them to find other living creatures, particularly birds and small mammals whose habitats were subtly integrated into their urban neighborhood. Insects were, he noted, particularly available and quite interesting once children became interested in the details of their anatomy and their habits. The investigation was to be non-intrusive. Hodge had a strong conservationist outlook that he couched in biblical language; an entire section in his first chapter contained quotations from Genesis and discussed human responsibility for "dominion over the animals." In *Nature-Study Review*, for example, he urged readers to watch for passenger pigeons, holding out the hope that these once plentiful birds were not extinct, and offering a reminder that humans had a powerful impact on the animal world.[31]

The often contentious Hodge was highly opinionated and could be dismissive of nature study colleagues who pursued what he described as "knot-hole science" rather than positioning their teaching in the midst of local experience. Investigating natural objects and specimens without a sense of their place in the world or their use seemed to him narrow and irrelevant. In this, he agreed with Liberty Hyde Bailey that the traditional object-lesson method had proven inadequate because "it usually takes the objects out of their setting and thereby destroys their meaning; and, moreover, it develops merely the observational power."[32] Hodge went much further, however, and he wanted to factor in human action as part of the natural world. Nature study needed to recognize humans as part of the natural environment, shaping and being shaped by it.

Worcester was a factory town, and its population at the turn of the century was heavily immigrant, with crowded slum areas and considerable poverty and ill health. Two of Hodge's best-known projects, done in conjunction with the city schools, addressed immediate and contemporary issues. For the first, like Scott, he chose a familiar speciman that probably was typically viewed with repugnance and ignored as a subject for pupil study, namely, the mosquito. He pointed it out as one of the "commonest and often the most annoying insects we have," but that familiarity made it even more important to introduce in a systematic way.[33] Students were assigned to find mosquito larvae on "egg rafts" on standing water in the neighborhood and bring them to the school to be placed in aquariums, where their development could be analyzed closely. Two aquariums simulated natural environments; one held water with a single fish, and the other, with a bit of mud and a water basin, was home for a frog and covered with gauze to prevent the mosquitoes from flying away. In both cases, pupils could watch how these natural predators handled the emerging wigglers. At the same time, teachers were to introduce the topic of malaria, asking which pupils had had the disease or knew someone who had it and explaining how mosquitoes carried the disease. The final lesson was to intro-duce another gauze-covered aquarium with the just the egg rafts and have the students watch them molt, pass through their larval stage, and then begin to emerge as adults. Pupils then put kerosene on the water, watched the wrigglers rise to the top, touch the oil, only to drop back dead, while the surviving adult mosquitoes tried to walk on the oil but sank into it (although they had been able to walk earlier on the water surface held in tension). Here was a highly effective way to eliminate mosquitoes. Hodge had the students take this knowledge into their neighborhoods, look for standing water, and tackle the problem by using kerosene or another oil.[34] His nature study lesson thus extended into and edu-cated the community through the children, and it demonstrated that careful observation could lead to understanding the dynamics of nature and have prac-tical outcomes.

Not all projects required that students work out-of-doors, and Hodge used the classroom itself as an unexpected laboratory to replicate biological research results.[35] In another extensive project, he created a "hygiene brigade" of pupils that tested out the notions about germs. Louis Pasteur's name had became a household term and discussions of bacteria common, but the public was not always clear about how to make use of this knowledge. Hodge taught teachers how to use the microscope to study these "animals" with their students and discuss the dangers they posed to human health. He directed a group of pupils in one classroom (all girls) to use damp washcloths to clean all the desks and surfaces each day (figure 14). In a second classroom, a similar group used dusters, which moved around the accumulated dirt and grime. He had the pupils in each room keep a record of the number of pupils who missed school because of colds during the course of the year. At the end of the year, they graphed the number of pupil days lost to illness in both classrooms to demonstrate that the more thorough cleaning had essentially eliminated the germs that led to colds. Other projects were more prosaic, including a competition to grow strong flower plants or to raise moth larvae through their developmental stages. All these projects were presented as ways to learn something of use, including the aesthetic value of flowers or the elimination of moth damage to carpets and clothing.[36]

Hodge's text covered botany, birds, and other topics, typically with a perspective that linked, as his title suggested, the deliberate content of nature study and pupils' immediate human life experiences at level they could compre-hend. Otis Caldwell's warm review of the book, although skeptical about the recommendation to use legend and myth, stressed Hodge's social and economic

Figure 14. The Health Brigade (all girls) of Upsala School, Worcester, Massachusetts, participated in an experiment to demonstrate that the elimination of germs by careful cleaning of desks and other classroom objects led to fewer colds. The photo is from Clifton F. Hodge, *Nature Study and Life* (1903), 476.

approach, along with the book's very accurate information.[37] Printed by the sixth-largest U.S. textbook publisher, Ginn and Company, *Nature Study and Life* was widely distributed in the United States and Britain. With Jean Dawson, who had spent two years as a fellow at Clark University, Hodge subsequently produced a high school biology book entitled *Civic Biology*.[38] Dawson, who taught at Cleveland Normal School, had also experimented with ways to present biology in the context of life lessons. She wrote an engaging fictional account of a school experiment where students playacted civic assignments, and her account follows the pupil assigned to be superintendent of public health as he promoted school gardens, cleaned up street waste, and informally lectured on the dangers of tetanus.[39] Hodge and Dawson represented the end of the spectrum in nature study that used the rhetoric of practicality and emphasized the ways biology was closely connected to social and civic responsibility, while also including topics that were becoming staples in high school teaching, including evolution, natural selection, and genetics.[40] These texts, like those of Scott, provided demonstration projects designed to encourage teachers to think creatively about the ways they could use local circumstances, often in urban or semi-urban areas, to engage pupils in learning more about science in their immediate worlds. The technique and subjects for teaching nature study in rural areas posed quite specific challenges and opportunities.

Anna Botsford Comstock shared with Hodge a sensibility about teaching pupils to be responsible, but her attention was on teachers in small, one- and two-room rural schoolhouses, typically women whose preparation and circumstances were dramatically different. Upstate New York was suffering from a declining economy, as farmlands were exhausted and many farm families were moving to the already crowded metropolitan areas where unemployment was high. The legislature, encouraged by the Hatch Act, had invested in the New York State Agricultural Experiment Station in Geneva, New York, and the College of Agriculture affiliated with Cornell University. Working at the young university with other faculty whose primarily responsibility was research and extension programs farm men and women, Comstock urged her colleagues to think broadly about their charge to improve rural life and prepare the younger generation through education. She began programs for rural and small-town teachers, whose lives she understood because she had spent some time teaching before taking a bachelor of science degree in biology at Cornell and marrying professor John Comstock.[41] With Liberty Hyde Bailey, later dean of the college, she envisioned nature study as something that lifted pupils above the routines of farm life and enabled them to see and appreciate their semi-wild surroundings. Between 1895 and 1905, Cornell faculty members produced numerous materials to assist teachers, taught in summer and weekend teacher institutes, and brought teachers to campus to learn nature study methods. Their series of leaflets, which became the model for state programs from Maine to California, were widely distributed in New York. Like Hodge's projects, they were intended to both equip teachers with information and provide a model of work that could be done with pupils. Short—twelve to sixteen pages—topical, and illustrated, these little brochures were distributed free and gave high visibility to Cornell's summer school and correspondence courses (figure 15). The staff also worked directly with schoolchildren across the state by coordinating Junior Naturalist Clubs, which were among the precursors to 4-H.

Building on leaflets she had written and additional material, Comstock compiled a substantial *Handbook of Nature Study*.[42] Organized in four parts—the teaching of nature study, animals, plants, and earth and sky—the volume concentrated on naturally occurring topics that would be familiar to pupils in the semirural landscape of Upstate New York and beyond.[43] The six-hundred-plus page manuscript proved dauntingly large for state and commercial publishers, who turned it down, so, rather reluctantly, the firm her husband and a colleague had established in Ithaca produced it. To John Comstock's surprise, the book became a leading income producer for the press, dramatically outpacing his scholarly books on insects. The publishing firm had as its motto, "through books to nature," a phrase that captured the sentiment behind the Comstocks'

The toad in various stages of development from the egg to the adult.

Figure 15. In the best of nature study, realistic detail could combine with abstraction to intrigue young minds, as in this illustration by C. W. Furlong. The illustration is from Simon H. Gage, "Life History of the Toad," *Nature-Study Leaflets* 14 (Albany, N.Y.: J. B. Lyon, 1904), 185.

publications in that the literature provided techniques for approaching the natural world.

Comstock's goal was to put children "in sympathy" with nature and stimulate their imaginations. An artist in her own right, she encouraged teachers to emphasize visual materials. Students were to compile and graph records of local weather, draw the environmental surroundings of a brook, and attend to the changing colors of local wildlife through the seasons and from youth to older age. Her aim was to teach them to be observant about the ordinary life around them in detail and in context, yielding insights more informative than simply describing a single specimen. She would probably not have agreed with Samuel Schmucker of Westchester (Pennsylvania) State Normal School, who suggested, "Accurate information gained from a book is not more valuable than slipshod information taken from personal observation."[44] However, she certainly did recognize his contention that "there is a joy in its acquisition and a clearness in the result, which make such information a far more lasting part of one's general store than is information gathered from a book." Indeed, her husband John had written of his own experience in the field that he "had never yet taken a class [of college students] into the field without finding something new."[45] This emphasis on a sense of discovery was important, but Anna was no sentimentalist when it came to observations. Witty and opinionated, she urged pupils to listen carefully and distinguish the communication sounds and patterns of their chickens, for example, and she showed little enthusiasm for mechanical incubators that produced "imbecile" offspring with no capacity for social interaction. She also appealed to children's visual imaginations and both produced finely detailed woodcuts herself and drew in other artists to illustrate her volumes and various Cornell publications on nature study.

Comstock worked directly with teachers. She offered a correspondence course taken by about two thousand of them, provided lectures to tens of thousands at numerous weekend institutes across the state, and supervised well-attended summer programs to acquaint them with both the materials and pedagogy of nature study.[46] Her constant message was that these teachers, most of them women, should offer more than facts memorized by rote. They had the capacity to use knowledge of nature to shape the moral outlook of their students, including respect for living things. Their goal should not be to transfer information but to build on the interests and curiosity of children, taking advantage of their imagination and experience. Her published essays about particular animals and plants were intended, she suggested, to help the pupils see relationships in nature. Comstock's handbook offered content, techniques (but little theory), and an approach intended to reinforce teachers' confidence. She recognized that it took courage to allow pupils to ask open-ended questions, some of which were beyond the competence of a teacher, but she reminded educators that they could thus share their students' excitement of asking questions and discovering answers. At this stage, turning to the knowledge of experts in books to correct, confirm, or extend observations was appropriate.

Scott, Hodge, and Comstock all wrote textbooks generously sprinkled with photographs, sketches, and other illustrations.[47] These authors sought to inspire teachers with a sense of possibility for teaching well beyond their classrooms, so their examples and advice were meant to be transparent, flashing images through which teachers looked to nature itself. That is, teachers were to bring their training and knowledge to the objects, but only as these could enhance the activities of their pupils. Nature study textbooks were typically a deliberate turn from reliance on expertise in the first instance and a corrective to more didactic manuals and self-help books in botany, ornithology, and other subjects meant for adults. Textbooks for teachers and teachers in training emphasized the importance of knowing (or knowing how to find) facts and ideas of experts, but this information was only a backdrop, since children needed to explore nature and describe it in age-appropriate ways. Pupils' genuine learning (and indeed enthusiasm for learning) was facilitated by their sense of discovery, based on their personal observations. In turn, that experience established a personal framework for information that pupils would retain and test in future years.

The nature study textbooks, part of a new, more self-conscious genre of pedagogy books, were distinct from older manuals as well as from the rather thinly descriptive "courses of instruction" distributed by urban school administrators with simply an outline of topics to be covered simultaneously throughout the school systems.[48] The texts for teachers were similar, however, in linking the subject to pedagogical theory even as they argued that nature study, with its emphasis on local investigation, needed to be designed for the particular landscapes accessible to the pupils. Teachers could choose among the dozen or more leading textbooks that offered more or less visual material, provided details about subjects readily available to them, and used methods that seemed effective. The loose philosophy that underlined the nature study movement emphasized direct contact with nature by pupils who learned to observe and reflect on their observations. Scientists, administrators, textbook writers, and teachers wanted one core outcome to be pupils who had learned to how to learn.

The three books discussed in detail in this essay were written by enthusiastic nature study advocates who argued for hands-on pedagogy even as they sought to guide inexperienced teachers to feel comfortable with subjects drawn from life and physical sciences. They were distinctive in style and content even as they presented similar subjects. They varied somewhat in their intended audiences, with authors well aware that nature study was implemented in rural, urban, and new suburban schools and in environments as diverse as the rocky coast of Maine; the prairies of Nebraska; or the flat, dry climate of central California. Hodge sought to encourage teachers in urban schools while Scott addressed those in small towns and cities; Comstock had all those readers in mind but added material for teachers in rural and farming areas. The variety of texts seemed to Liberty Hyde Bailey a good thing. He encouraged teachers to "find one to your liking," and he suggested that newer books offered good ideas but that they need not replace the older ones. In fact, he pointed out, "when one textbook satisfied everybody, it is because everybody is uncritical and unpersonal."[49]

Clifton Hodge went so far as to suggest that nature study be at the core of the entire elementary curriculum. He intended his textbook as an aid to teachers and intended his material to lead away from books to active learning. He wrote, "Instead of giving over our entire school system to passive book learning, we shall have at least one subject that shall keep alive in the child the spirit of research, under the impetus of which he makes such astounding progress in learning the great unknown of nature around him during the first three or four years of life."[50] His book is replete with potential topics and specific examples, all of which he intended as evocative information a teacher might use to respond to the objects brought to class. The model of the anecdotal examples was influential and spurred many teachers to report on their successful projects of ant farms, bird-feeder placement, and unusual excursions in the pages of the *Nature-Study Review*, once that specialized journal was established, in 1906, and at professional meetings.

Comstock and her Cornell colleagues, less steeped in pedagogical and psychological theory than Scott and Hodge, also wanted the nature experience to be central but did not resist print. Their Junior Naturalist Club members, for example, were encouraged to keep journals, and their observations were taken seriously as points of departure for discussion and resolution. The written word was important even for young readers, but reading was complementary to their observational studies. Comstock's *Handbook* was directed toward a somewhat ambiguous audience; it was primarily accessible to teachers but often slipped into language that anticipated advanced students. Her persisting production of classroom materials suggested that she did think it was possible to go "through books to nature." Teachers simply needed to be alert to the danger of relying too heavily on printed material, which would make "the work too complete, too rigid, and too formidable."[51]

One outcome of the emphasis on the child rather than the content was that scientists became less comfortable with nature study because it seemed in opposition to the positivist mode that gained strength through the 1930s. Academic experts were much more comfortable with a new model of elementary science introduced in the early 1930s that depended on crafted readers that taught materials which seemed to indeed be science at an elementary level and with clear reference to the steady advancement of knowledge up through the now graded schools.[52] Nature study was just one element of the progressive schoolroom that faced challenges during the social conservative swing in the 1920s; administrative women who supervised nature study did as well.[53] Over time, the nature study emphasis on tactile experience retreated to a place supplementary to elementary science texts, but its successful hands-on methods proved durable into the twenty-first century.

The long echo of the injunction to "study nature, not books" had resonated throughout the nineteenth century. For transcendentalist Ralph Waldo

Emerson, it was a call to acknowledge the highly personal way individuals might relate to the natural world and a reaction to expertise that seemed narrow and static. For zoologist Louis Agassiz, who made each new Harvard zoology student spend hours studying and then describing a single fish, it meant showing appropriate skepticism toward older assumptions about specimens and their relationships to one another. Both men, of course, wrote books and referenced the work of others, but they insisted that nature itself was in some sense to be among the teachers. They and others were reacting to unquestioning accept-ance, even dependence on other experts and wanted instead individual and independent thinking. When the nature study educators issued a similar call, they were reacting to rote learning and recitation, often their own negative experience in school. As Liberty Hyde Bailey recalled, "We were taught botany by using "Gray's 'How Plants Grow' as a reading book, without a word of explanation, with absolute silence on the 'teacher's' part and incomprehension on ours."[54] The new curriculum thus had to negotiate its way through the issue of print culture, where access to expertise did not automatically lead to education.

Nature study advocates wrote books, illustrated them, and encouraged their use by a generation of educators not previously exposed to such studies. Inevitably, their introductions articulated the mission of nature study, which was to "open the pupil's mind by direct observation to a knowledge of the common things and experiences in the child's life and environment."[55] At the same time, the authors emphasized that books were complementary to direct contact with the natural world, sometimes a stimulus to engage in such studies, sometimes a way to discover how personal knowledge fit with more established wisdom, and sometimes a contribution to enhancing fundamental understanding of a phenomenon. The focus of nature study was nature itself, but "not books" served only as a warning that books were insufficient in the goal to establish a personal, environmental, and sense-based understanding. In a subtle but signifi-cant way, the nature study educators believed that objects observed in familiar settings were intellectually fungible and that they opened out to multiple meanings with intellectual, social, aesthetic, and emotional implications.[56] Books offered a way to draw pupils back to the vocabulary and standard descrip-tions found in their community and, eventually, among experts.

Nature study texts proved indispensable to the success of nature study in the public schools. They were used extensively in teacher training and by class-room teachers. This was a moment in the history of American education when there seemed considerable confidence that well-guided teachers had the capacity to develop individual projects that utilized their particular environment and addressed the interests and needs of their pupils. Teachers, like their pupils, responded to the study of natural elements when these were presented in a wide range of contexts and with appreciation for their complexity. Written sources and the opinions of experts were valued but used with caution in elementary education. Disclaimers were common in the prefaces to the textbooks and readers used by teachers and their pupils. Anna Botsford Comstock, for example, argued that the use of readers should be restricted and their stories not read until "after the pupils have completed their own observations on the subjects of the stories."[57] Perhaps because literacy was presumed to be achieved, or at least achievable, educators were positioned to think about what other abilities and qualities they hoped to instill in their pupils.

Books were and remained essential to studies of the natural world. Nature study methods, however, had posed a challenge about the ways printed textbooks could and should be used in conjunction with pedagogy. Nature study was in some measure a reaction to rote learning and a challenge to the book culture on which such education relied. Its adherents provided a new style of books to teach something of both content and pedagogy. The methods of this new curriculum, outlined in teachers' manuals, presented techniques for unleashing curiosity of pupils, constrained primarily by the insistence on careful observation and shared insights with their peers. The guidelines were open-ended, as pupils and teachers explored local environments. Perhaps not surprisingly, the learning outcomes could be ambiguous and not easily measurable by increasingly technical grading methods. A changing outlook in the next generation of educators led them to push for standardized textbooks in carefully graded classrooms.[58] For the generation of nature study authors like Scott, Hodge, and Comstock, however, the enthusiasm of many teachers and pupils seemed sufficient proof that their methods were effective. The stimulus of detailed examples and often inspirational rhetoric in books made a major contribution to original, stimulating, and educational projects at the heart of nature study.

Notes

1. Agassiz may or may not have explicitly stated this, but it held connections to the Transcendentalist Ralph Waldo Emerson, and Agassiz's students frequently cited it. This heuristic reminded both his college students and those to whom he lectured at normal schools and his seaside summer school to inspect natural history specimens carefully before looking for information in books. See commentary by Nancy Pick, *Rarest of the Rare: Stories behind the Treasure of the Harvard Museum of Natural History* (New York: HarperResource, 2004), 16.

2. Thus Liberty Hyde Bailey made a distinction between the older imported system of "object lessons" and the use of specimens to teach nature study in his *Nature-Study Idea: Being an Interpretation of the New School Movement to Put the Child in Sympathy with Nature* (New York: Doubleday and Co., 1903), 5.

3. Charges of corruption and scandals relating to textbook competition and sales pushed city and state systems to adopt system-wide textbooks in the late nineteenth century, according to Joseph Moreau, *Schoolbook Nation: Conflicts over American History*

Textbooks from the Civil War to the Present (Ann Arbor: University of Michigan Press, 2003); also see Michael W. Apple, "Regulating the Text: The Socio-Historical Roots of State Control," in *Textbooks in American Society: Politics, Policy, and Pedagogy*, ed. Philip G. Altbach et al. (Albany: State University of New York Press, 1991).

4. On the emerging credibility of texts since the early modern period, see Adrian Johns's introduction to *The Nature of the Book: Printed Knowledge in the Making* (Chicago: University of Chicago Press, 2000), 1–2.

5. See the account by Samuel H. Scudder, "In the Laboratory with Agassiz," *Every Saturday* 16 (April 4, 1874): 369–70; Margaret Welch, *The Book of Nature, 1825–1875* (Boston: Northeastern University Press, 1998) stresses the role of print culture, texts and illustrations, as part of formal and informal natural history teaching and practice in mid-century North America (22, 134).

6. On scientists' early endorsement of nature study see Sally Gregory Kohlstedt, "Study Nature, Not Books: Scientists and the Origins of the Nature Study Movement," *Isis* 96 (September 2005): 324–52.

7. Christine A. Ogren, *The American Normal School: "An Instrument of Great Good"* (New York: Palgrave Macmillan, 2005). Ogren has a very useful appendix listing all the public normal schools, with information about their founding dates and changing names.

8. For Hall and the child study movement, see Dorothy Ross, *G. Stanley Hall: The Psychologist as Prophet* (Chicago: University of Chicago, 1972), chap. 15.

9. Barbara Beatty, "Psychologizing the Third R: Hall, Dewey, Thorndike and Progressive Era Ideas of the Learning and Teaching of Arithmetic," in *When Science Encounters the Child: Education, Parenting, and Child Welfare in Twentieth-Century America*, ed. Barbara Beatty, Emily D. Cahan, and Juliet Grant (New York: Teachers College Press, 2006).

10. Literacy was not, of course, equally distributed. Data shows illiteracy was at 4.6 percent for the native-born and 12.9 percent for the foreign-born white population; the non-white illiteracy rate was 44.5 percent. Carl Kaestle et al., *Literacy in the United States: Readers and Reading since 1880* (New Haven: Yale University Press, 1991).

11. Kathleen McDowell, "The Cultural Origins of Youth Services Librarianship, 1875–1900" (PhD diss., University of Illinois, 2007); Abigail A. Van Slyck, *Free to All: Carnegie Libraries & American Culture, 1890–1920* (Chicago: University of Chicago Press, 1995); and Jacalyn Eddy, *Bookwomen: Creating an Empire in Children's Book Publishing, 1919–1939* (Madison: University of Wisconsin Press, 2006). Also see the list of books titled "Juvenile Nature Literature and Nature Pedagogy" at http://lcweb2.10c.gov/ammem/amrvhtml/consbib10.html.

12. Charles B. Scott, *Nature Study and the Child* (Boston: D. C. Heath, 1900), 107. Scott taught semiweekly classes for more than 150 teachers through the St. Paul Teachers' Association in 1891 and 1892 in addition to his biology high school classes; he apparently also visited some of their schools to observe their classroom practices. J. M. Rice, *The Public School System of the United States* (New York: Century Co., 1893), 190–91.

13. Audrey B. Champagne and Leopold E. Klopfer, "Pioneers of Elementary School Science: Wilbur Samuel Jackman," *Science Education* 63 (1979): 145–75. Jackman had graduated from Southwestern Normal School in California, Pennsylvania, before going to Harvard.

14. Wilbur S. Jackman, *Nature-Study for the Common Schools* (New York: Henry Holt and Co., 1891). The text was comprehensive in its coverage, including not only botany

and zoology but also meteorology, geography, mineralogy, chemistry, and physics as these could be observed and related to the daily lives of pupils.

15. Wilbur S. Jackman, *Nature Study and Related Subjects for the Common Schools*, part 1, *Notes* (Chicago: The Author, 1896), 29.

16. Ibid., 12.

17. Some recent historical work shows the possibilities of such analysis, including Susan Schulten, *The Geographical Imagination in America, 1880–1950* (Chicago: University of Chicago Press, 2001), and, at the collegiate level, David Kaiser, *Drawing Theories Apart: The Dispersion of Feynman Diagrams in Postwar Physics* (Chicago: University of Chicago Press, 2005).

18. "Nature Study in Our Schools," *Denver Daily News*, March 3, 1895, 12.

19. Barbara T. Gates, *Kindred Nature: Victorian and Edwardian Women Embrace the Living World* (Chicago: University of Chicago Press, 1998), 53. In the eighteenth century, educators had debated whether fables featuring animals were inappropriate, and Jean-Jacques Rousseau had harshly critiqued this literature; see Louise E. Robbins, *Elephant Slaves and Pampered Parrots: Exotic Animals in Eighteenth-Century Paris* (Baltimore: Johns Hopkins University Press, 2002), 160–64. On American nature study, see Kimberly Perez, "Fancy and Imagination: Cultivating Sympathy and Envisioning the Natural World for the Child" (PhD diss., University of Oklahoma, 2006).

20. John Tebbel, *A History of Book Publishing in the United States*, vol. 2, *The Expansion of the Industry, 1865–1919* (1972; repr., New York: R. R. Bowker Co., 1981), 574. Eric Lupfer, "Reading Nature Writing: Houghton Mifflin Company, the Ohio Teachers' Reading Circle, and *In American Fields and Forests* (1909)," *Harvard Library Bulletin* 13 (Spring 2002): 37–58.

21. Edward F. Bigelow, *How Nature Study Should Be Taught* (New York: Hinds, Noble & Eldredge, 1904); he was citing essayist Hamilton Mabie.

22. Anna E. McGovern, *Nature Study and Related Literature, Showing Aims and Plans of Teaching with Illustrative Lessons for the First Four Grades* (Cedar Rapids, Iowa: Republican Printing Co., 1902); McGovern taught at Iowa State Normal School, later Northern Illinois University.

23. Scott, *Nature Study and the Child*, 329. Scott's introduction discusses this project and his simultaneous role as supervisor of science work in the elementary schools of Oswego from 1894 to 1899. Scott, a graduate of the University of Michigan, ended his career in Puerto Rico, where he helped establish a normal school; he died of typhoid fever in Plainfield, New Jersey, in 1904. *Oswego (N.Y.) Daily Times*, June 21, 1904, 4.

24. Charles B. Scott, *Nature Study in the Elementary Schools* (Concord, N.H.: Republican Press Association; reprinted from *Proceedings* of the American Institute of Instruction, 1894). Scott also addressed literally thousands of teachers at the National Education Association meeting in Asbury Park; see "The Country Schoolhouse," *New York Times*, July 11, 1894, 8.

25. Herald B. Dunkel, *Herbart and Herbartianism: An Educational Ghost Story* (Chicago: University of Chicago Press, 1970).

26. See, for example, the series of four *Nature-Study Readers* written by John W. Troeger (some with Edna Beatrice Troeger). All were written in the first person, from the point of view of an adventuring Harold, including *Harold's First Discoveries, Harold's Rambles, Harold's Quests, Harold's Explorations*, and *Harold's Discussions*; published by D. Appleton and Company between 1897 and 1902.

27. Charles B. Scott, *Nature Study and the Child* (Boston: D. C. Heath, 1900), 522,325.

28. Scott's textbook, *Nature Study and the Child*, was also reprinted in at least one English edition, published by Isbister in 1901.

29. Wilbur Jackman continued to produce nature materials and published some of his texts with both Holt and Macmillan, sometimes as reprints, but the lists of books most often used by teachers after the turn of the century rarely listed his textbooks. His *Nature Study for the Common Grades: A Manual for Teachers and Pupils below the High School in the Study of Nature* (New York: Macmillan, 1909) was republished after his death, having been initially released a decade earlier.

30. Clifton Hodge, *Nature Study and Life* (Boston: Ginn and Co., 1903). It was listed with the "out-of-door" books in the *New York Times Book Review*, April 19, 1902, 15.

31. Clifton Hodge, "Passenger Pigeon Investigation," *Nature-Study Review* 6 (May 1910): 110–11.

32. L. H. Bailey, *The Outlook to Nature* (New York: Macmillan Co., 1905), 217.

33. Hodge, *Nature Study and Life*, 62. Hodge was a fellow at Clark, 1889–91; assistant professor of physiology and neurology, 1892–1906; and professor of biology, 1906–13. In the latter year, he moved to the University of Washington as professor of social biology, and by 1919 he was professor in the General Extension Division of the University of Florida. His experiments with domesticating wild birds and his "war against cats in Worcester," which got considerable attention, reflected his somewhat controversial stances; see "Turning the Partridge into a Quiet Barnyard Fowl," *New York Times Sunday Magazine*, November 26, 1905, 3.

34. Clifton C. Hodge, "Practical Work with Mosquitoes," *Nature Study Review* 3 (1907): 33–36.

35. On the public interest around issues of germs and sanitation, see Nancy Tomes, *The Gospel of Germs: Men, Women and the Microbe in American Life* (Cambridge: Harvard University Press, 1998).

36. Hodge, *Nature Study and Life*.

37. *School Review* 10 (November 1902): 114–15. Like others, Caldwell was intrigued by Hodge's description of nature study as "learning those things in nature that are best worth knowing to the end of doing those things that make live most worth living."

38. Clifton Hodge and Jean Dawson, *Civic Biology: A Textbook of Problems, Local and National, That Can Be Solved Only by Civic Cooperation* (Boston: Ginn and Co., 1918).

39. Jean Dawson in *The Boys and Girls of Garden City* (Boston: Ginn and Co., 1914) presented a set of examples about the ways biology could be introduced in the context of civic problem solving. Dawson received a PhD (1905) from the University of Michigan; she then spent two years as a fellow at Clark (1905–07) before teaching in the Cleveland Normal School, and was active in helping develop an anti-fly campaign, including an educational film that garnered international attention. See *New York Times Sunday Magazine*, May 19, 1912, 11.

40. See chapter 7 on high school biology in Philip J. Pauly's *Biologists and the Promise of American Life: From Meriwether Lewis to Alfred Kinsey* (Princeton: Princeton University Press, 2000), and Robert A. Hellman, "Evolution in American School Biology Books from the Late Nineteenth Century until the 1930s," *Biology Teacher* 27 (December 1965): 778–80.

41. Anna Botsford Comstock, *The Comstocks of Cornell: John Henry Comstock and Anna Botsford Comstock* (Ithaca, N.Y.: Cornell University Press, 1953).

42. Anna Botsford Comstock, *Handbook of Nature-Study* (Ithaca, N.Y.: Comstock, Press, 1911). Comstock's *Handbook* has remained in print ever since its first publication. In his review, Vernon Kellogg found it encyclopedic and thought, accurately, that it would become "the book of American nature study." See *Science* 37 (May 2, 1913): 669–70.

43. A good discussion of the borderlands that had been settled and mapped in a preliminary way but were not necessarily in continuous use by humans is found in Robert E. Kohler's *All Creatures: Naturalists, Collectors, and Biodiversity, 1850–1950* (Princeton, N.J.: Princeton University Press, 2006).

44. Samuel Christian Schmucker, *The Study of Nature* (Philadelphia: J. B. Lippincott Co., 1908), 22.

45. John Henry Comstock with Anna Botsford Comstock, *Insect Life: An Introduction to Nature-Study and Guide for Teachers Interested in Out-of-Door Life* (New York: D. Appleton and Co., 1899), 7.

46. Her autobiographical *The Comstocks of Cornell* remains the best account of her activities.

47. Visuals were an important component, offering a way to see nature when the specimens under discussion were not available. Nature study educators thus drew on a long tradition of illustration in natural history, particularly strong after the seventeenth century. Publishers produced such materials for classroom use, such as Putnam's several series of Nature Study Pictures, widely advertised; see *Publishers Weekly* 54 (July 9, 1898): 89.

48. A significant number of these ephemeral course materials have been collected and are in the files of the Teachers College Archives, Gottesman Libraries, Columbia University. Bibliographies of textbooks and of related reading on insects, birds, plant life, trees, rocks and minerals, and physical nature study were common, found in these courses of study and also published by state and federal governments; see, for example, James Ralph Jewell, *Agricultural Education, Including Nature Study and School Garden*, 2 (Washington, D.C.: Department of Interior, Bureau of Education, 1907).

49. Bailey, *Nature-Study Idea*, 198–99.

50. Ibid., 14.

51. Liberty Hyde Bailey, "Letter of Transmittal," *Cornell Nature-Study Leaflets* (Albany, N.Y.: J. B. Lyon Co., 1904), 2.

52. Kimberley Tolley, *The Science Education of American Girls: An American Perspective* (New York: Routledge Falmer, 2003), discusses Gerald Craig's role in emphasizing elementary science over nature study; his papers are at the Wisconsin Historical Society, Madison.

53. Sally Gregory Kohlstedt, *Teaching Children Science: Hands-on Nature Study in North America, 1890–1930* (Chicago: University of Chicago Press, 2010), 201–14.

54. Edward Bigelow, *The Spirit of Nature Study* (New York: A. S. Barnes, 1907), 145. Similarly, L. H. Bailey in *The Nature Study Idea* wrote, "Nature-study is a revolt from the teaching of formal science in the elementary grades" (5).

55. Bailey, *Nature-Study Idea*, 4.

56. This recognition of the use of objects in classrooms is somewhat resonant of the border objects in museums as discussed by Susan Leigh Star and James R. Griesemer, "Institutional Ecology, 'Translations' and Boundary Objects: Amateurs and Professionals

in Berkeley's Museum of Vertebrate Zoology, 1907–1939," *Social Studies of Science* 19 (1989): 387–420.

57. Anna Botsford Comstock, "Suggestions for Nature Study Work," *Cornell Nature-Study Leaflets* 5 (Albany, N.Y.: J. B. Lyon Co., 1904), 55.

58. Gerald Craig, a one-time Nature-Study Society member, was among the most productive of the new textbook writers and his book series for Ginn and Company, *Pathways in Science*, assumed a kind of universal knowledge that could be calibrated by grade.

Basic Seven, Basic Four, Mary Mutton, and a Pyramid

The Ideology of Meat in Print Culture

RIMA D. APPLE

American love of meat has been and continues to be renowned. Nutritionists could and did claim that so much meat was unhealthful or unnecessary and advised that families eat less meat. Simultaneously, nutritional scientists also urged people to eat meat for their health, both physical and mental. Meat was highly controversial; it was very popular. In the drive to educate Americans, primarily female consumers, about the nutritional value of meat in a healthful diet, we see the power of print culture, of text and graphics, to shape the dimensions of the discussion. The ideology of meat—the belief that meat is the indispensable food in the American diet—rests on its scientific and medical image, an image that permeates and dominates twentieth-century popular and educational literature directed at women and children. Articles published in newspapers and the growing number of women's magazines in the United States, as well as government pamphlets and reports, placed the welfare of the family firmly on the shoulders of the mother, who was expected to utilize contemporary science to provide healthful meals.[1] This chapter focuses on popular nutrition literature, particularly the power of government-produced print culture, to demonstrate the essence and stability of the authority of the ideology of meat in our society.

By the twentieth century, federal and state agencies had recognized that print culture played an especially critical role in bringing the advantages of modern science to the American populace. As early as the 1890s, alongside bulletins on scientific agriculture, the U.S. Department of Agriculture (USDA) produced pamphlets describing contemporary nutritional findings. Most significant for this essay were the publications directed toward the mothers and housewives of the country. The earliest bulletins, such as W. O. Atwater's 1895 *Methods and Results of Investigations on the Chemistry and Economy of Food*, were more technical than later ones, but they all aimed to help women apply contemporary science to their domestic tasks.[2] Beginning in the 1900s, the cooperative extension services of states such as Wisconsin and New York used popular brochures to instruct women about healthfully feeding their families. Caroline L. Hunt, the first director of home economics at the University of Wisconsin, understood the potential of print culture to disseminate useful science to a broad audience. Under her aegis, the first university home-economics publication detailed experiments conducted by her students on the efficacy of the "fireless cooker," a precursor of today's slower cooker.[3] Hunt resigned from her position in 1908 and later joined the USDA in its Bureau of Home Economics, from which she issued the bulletins that formed the basis of the widely respected and frequently reproduced federal guidelines.[4] This was print culture very consciously aimed to educate the individual woman, the individual mother, with the most up-to-date scientific information she needed to fulfill her role as protector of the family's welfare, if necessary to change her commonsense beliefs to more reflect contemporary science.

Women's Roles in Nutrition

From at least the nineteenth century, women have been admonished to closely monitor their families' diets. In the widely read *Ladies' Home Journal*, commentators such as Sarah Tyson Rorer, leading nutritionist of the early twentieth century, insisted that unless a woman takes this responsibility seriously, she "is not a true woman."[5] In the home-economics journal *Forecast*, another columnist painted a bleaker picture, linking poor cooking to a rise in crime. The logic was simple: bad meals drive the man of the house to drink, and from drink to criminal activity is one short step.[6] Frequently, women's dietary obligations reached beyond the home to the nation at large, as one educator asserted in 1903: "Food is the basis upon which physical life is supported. The intellectual, and to some degree, the moral vigor is measured by the physical state. . . . The history in general of the individual and of a nation where properly selected and prepared foods have been habitually used, is that of moral, intellectual and physical strength."[7] In a woman's kitchen rested the fate of the country. Over time, the rhetoric became less picturesque, but the point remained. Women needed nutrition education in order to protect their families and society. If children struggled in school, if husbands were not productive at work, if families were not successful, the woman had failed in her responsibility to provide an appropriate diet.[8]

Nutritionists, doctors, teachers, food manufacturers, and reformers all directed their instruction to literate, middle-class women. But writers also reminded readers that "money is not so important as intelligence when it comes to providing an adequate diet for children," explicitly associating intelligence with reading and following the latest nutrition advice.[9] Similarly, educators anticipated that children hearing contemporary nutritional standards in school would carry that information back to their homes for the edification of their families.[10] Books, magazines, advertisements, government pamphlets, and exhibitions all employed the language of science—whether it came from the laboratory, the field, or the copy writer's pen—to educate women about healthful foods. "Do you know," wrote one Cornell University home economist in 1949, "that the food you [mothers and future homemakers] choose can make a difference in the growth and development of your children? That the food you select can make a difference in the number of active years you and your husband will have? That the right food may prolong the vigor of life?"[11] Woman's role as wife and mother, as housekeeper and guardian of her family's health and well-being, legitimated the growth of the burgeoning domestic advice-literature industry of the late nineteenth and twentieth centuries, a medium that documents the ideology of meat.[12]

What Is Meat?

Print culture proclaimed the importance of meat, without any consensus on its definition. A 1905 Cornell University pamphlet for "farmers' wives" listed beef, mutton, pork, veal, and fish as meat.[13] A home-economics textbook considered meat "the flesh of animals used for food," expanding the scope of the word: "In this country the meats in ordinary use are; beef and veal, obtained from cattle; pork, the flesh of the hog; mutton, obtained from sheep; fish; poultry, which includes chickens, ducks, geese, turkeys, the guinea fowl, and young pigeons, or squabs. Another meat is game, which is the flesh of wild animals and fowls, such as venison (the flesh of the deer), the rabbit, squirrel, quail, dove, plover, duck, and turkey."[14] Frequently, the word meat was not specifically defined at all.

A 1985 *New York Times* article by Jane Brody illustrates how slippery the definition of meat could be. Brody wrote that Americans had significantly reduced their meat consumption from 94.4 pounds per capita in 1976 to 78.8 pounds in 1983. She also noted that the consumption of poultry nearly doubled from 29.8 pounds to 65.9 pounds. Meat down, poultry up; so the decreased meat was red meat? Perhaps. But it could include pork, or lamb, or other animal protein, excluding poultry.[15] Government reports, which strive for precision in their analyses, highlight the ambiguous definition of meat. For example, when the Economic Research Service of the USDA studied meat consumption between 1970 and 1997, the agency writers concluded that "red meat (beef, pork, lamb, and veal) accounted for 58 percent of the total meat supply in 1997, compared with 74 percent in 1970. By 1997, chicken and turkey accounted for 34 percent of the total meat consumed, up from 19 percent in 1970. Fish and shell fish accounted for 8 percent of total meat consumption in 1997 and 7 percent in 1970."[16] In other words, pork was "red meat," though in these decades pork producers often advertised their product as "the other white meat."[17] Chicken and turkey were not red meat but were meat, as were fish and shellfish. Throughout the twentieth century and into the present, the definition of meat has remained problematic.[18] For purposes of this chapter, I use the broadest definition.

Studies of Meat Consumption

This semantic confusion prefigured problems with studies of meat consumption. Anecdotal evidence and sketchy surveys from the nineteenth century indicate that working-class families, as well as families of the middle and upper classes, commonly served "meat" two or three times a day. In 1915, home economist Ida Cogswell Bailey-Allen chided the middle-class readers of *Good Housekeeping* for their overdependence on meat, which she blamed on a lack of education. "The custom of serving meat three times a day is a common error," she explained, advising that "with such wide variety of protein foods from which to select, there should be no difficulty planning menus in which meat figures but once a day."[19] The World War I slogan "Do not buy a pound of meat until you have bought three quarts of milk" points to attempts to conserve meat and encourage the production of milk. It also suggests that despite rising meat prices and the equation of patriotism with meatless meals, at least one home economist despaired that "meat consumption goes on in many quarters at much the usual rate."[20]

More detailed studies from the early twentieth century were not any more precise. For example, in 1920, the Columbia University chemist Henry C. Sherman estimated that 32.7 percent of the "bill of fare of the average American" was meat, including poultry and fish, though the sample used in this study is not clear.[21] Other research of the same period compared the foods purchased by five hundred families with that bought by four hundred families "about twenty years ago" and concluded that Americans were eating 8 percent less meat than they had been two decades earlier, never defining meat or specifying the families surveyed.[22] What did the category "meat" include in this study? Who were the families? These questions make it impossible to evaluate these claims. In a different type of analysis in the 1940s, the National Restaurant Association reported America's top dishes as ham and eggs, prime ribs, chicken, lobster, New England boiled dinner, friend oysters, baked Virginia ham, breast of capon, filet of sole, and deviled crab, or, in the words of two historians of vegetarianism, "a meat lover's dream." Given the source of the study and the names of the meals, these appear to be favorite restaurant orders, nonetheless underscoring the ideology of meat.[23]

Later studies continued to be suggestive, not definitive. One government survey focusing on consumption during 1948 surmised that "among food groups, meat, poultry, and fish ranked first in the household food budget" and that the families surveyed spent an average of 29.5 percent of their food dollar on this category.[24] A 1963 text for nutrition educators casually announced that "in 1960 Americans appeared to consume 33 pounds more of red meat and poultry per person than in 1940" and that "since 1955, consumers have been eating more beef than pork per person."[25] By the 1970s, the U.S. Department of Agriculture calculated that "more than one-third of the home food dollar (value of purchased and nonpurchased food used) was spent on meat, poultry, fish, and eggs." This portion did not vary much over the year, but small seasonal variations led to higher beef consumption in the spring than the fall or winter, and pork consumption was highest in the spring and lowest in the summer.[26] In an attempt to be more specific, the Agricultural Research Service analyzed the data from the USDA Continuing Survey of Food Intakes by Individuals and Diet and Health Knowledge Survey, 1994–96. The researchers noted the mean number of meat group servings consumed per day across the survey's informants was 2.2. This contrasts sharply with the mean 1.3 servings from the milk group and 1.5 from the fruit group.[27] Evidently, regardless of income bracket, social or cultural characteristics, or time period, Americans ate a lot of meat.

Benefits to and Problems with Meat Eating

From the late nineteenth century, if not earlier, meat has dominated discussions of the U.S. diet, and advocates have found many reasons to celebrate its attributes. Women's and popular magazines were filled with articles urging women to feed their families meat two or even three times a day. This literature as well as the pamphlets distributed by the federal and state home-economics agencies published recipes centered on meat. Animal protein cured anemia; it encouraged the digestive juices and thus aided digestion; it complemented other foods and made them more attractive. Contemporary ethnographic studies confirmed the healthfulness of meat. The lives of Eskimos and of the explorers who lived with them provided evidence that all-meat diets were highly successful.[28] Meat contained high-quality protein to build strong muscles and was also a good source of thiamine, iron, and phosphorus.[29] Meat was an important element in the "gold standard of good nutrition," emphasized Hazel Kepler and Elizabeth Hessler in their 1950 book *Food for Little People*. Reminding mothers that "all fish, flesh, and fowl are bristling with protein and the B vitamins," the authors added a "consoling though [*sic*]; the cheaper, tougher cuts of meat are just as rich in these valuables, and are often more flavorful, than their more expensive relatives."[30] The iron content of meat was especially important for women's health. By the end of the twentieth century, nutritionists were particularly concerned that modern dieting crazes were encouraging women to cut back on meat to save calories. Magazines as diverse as *Redbook* and the *Saturday Evening Post* insisted that women "eat red meat, fish, or dark-meat poultry at least once a day" for its iron content.[31] The agricultural industry and food manufacturers joined this chorus promoting meat. Their advertisements in popular literature and in educational pamphlets further reinforced the ideology of meat.

A few decried what they saw as the overconsumption of meat in this country, and fewer still advocated vegetarianism.[32] Some were equivocal, using contemporary scientific claims to argue for and against meat consumption at the same time. One such cautious author, John C. Olsen, explained the crucial role that meat and protein played in sustaining human life: "The presence of protein in the human diet is of the highest importance, because the protoplasm or living portion of the human system can be built up and nourished only by this constituent of our food. . . . [T]he protein of meat is much more easily digested and more readily assimilated by the tissues than vegetable protein, so that meats are a very desirable constituent of the human diet." Protein is needed, and meat is the best source of protein, but, he added, "serious dangers attend the consumption of meats, especially in excessive quantities. This is due to the fact that . . . [proteins of meat] seem more liable to decompose in such a manner as to give rise to highly poisonous substances." Despite this warning, he recommended that each adult eat a half of a pound a day.[33]

Similarly, Mary Davis Rose Swartz, leading nutritionist of the turn of the twentieth century, recognized that meat was a vital component of the U.S. diet from the homemaker's perspective. It was relatively simple to prepare; it was an uncomplicated central point of a meal; and it contained a high percentage of protein; generally, "it is a valuable food, supplying nutriment in a pleasing and digestible form." Swartz accepted the health and aesthetic qualities of meat but also its potential dangers. For instance, when we eat too much meat, the body excretes the surplus through the kidneys, straining that organ. In addition, "purins" in meat cannot nourish the body. These are changed into uric acid, and, when not properly eliminated through the overworked kidneys, they can give rise to gout, rheumatism, and also putrefaction, leading to autointoxication.[34] Balancing all the benefits and all the difficulties, Olsen and Swartz advised Americans to eat less meat, though Swartz expected this would create a "domestic problem," namely how "to make palatable dishes" from dry peas and beans, peanut butter, and the like.[35] Bailey-Allen also suggested that women prepare less meat for their families: "When cooked, dried vegetables, including all kinds of beans, peas and lentils, are nearly equal to meat in nutritive value."[36] Cooked dried vegetables are "nearly equal," not equal. Meat is the standard, with substitutes something less than ideal. A mother concerned for the health of her children could easily read this as "Dried vegetables are possible. But to ensure a healthful diet, meat is best."

Others vehemently insisted on the primacy of meat in the human diet. In his 1935 *Eat, Drink, and Be Wary*, F. J. Schlink, founder of Consumers' Research, author of the best-seller *100,000,000 Guinea Pigs*, and caustic defender of consumer protection, took to task researchers who, he claimed, were in the service of the certain sectors of the food industry. He believed that this connection led scientists and home economists to highlight the role of bread and cereals. He singled out Swartz for her "fantastic theory" that "meat causes intestinal poisoning" and for ignoring that "the overuse of bread and cereals is . . . conducive to tooth decay." These assertions were based on experiments from the early twentieth century, he declared, and more recent laboratory findings superseded them. Furthermore, evolution proved that "man" required large quantities of meat. "For man and other carnivorous and omnivorous animals, the herbivorous animals, including cattle and swine, poultry, and of course fish, perform the valuable service of sorting and assembling from the more elementary substances consumed by insects, worms, larvæ, building up the essential amino acids from green plants, grain, seeds, hay, insects, etc., animalculæ, yeasts, molds, etc."[37] Schlink was a most flamboyant writer, and the popularity of Consumers' Research in the 1930s helped to spread his opinions widely.

A nutritional rival to meat arose with the discovery of vitamins in fruits and vegetables in the first third of the twentieth century. While Schlink disdained plant foods, others wrote in praise of these mysterious micronutrients in "corn, olive oil, and green stuff."[38] Vitamins, though, did not dislodge meat as the fundamental element in American cooking. Commentators feared that consumers would find a wholly vegetable diet "boring" and that people accustomed to eating animal protein two and three times a day would find it difficult to avoid all meat. Families needed meat for taste and because it was familiar, but the amount should be limited, some contended.[39] In the popular childcare magazine *Parents*, nutritionist Anne Pierce accepted that "eggs and milk and green vegetables may supplement a milk and cereal diet effectively as meats, theoretically, but meat has a savor, an appetizing value, and is tolerated and digested better in some cases than eggs."[40] Other foods are good, theoretically, but meat is what the family will eat most readily.

By the 1950s, meat's eight essential amino acids received increasing attention in the media. Discussion of these nutritional elements had appeared in popular literature since at least the 1920s, but with Adelle Davis's highly popular 1954 *Let's Eat Right to Keep Fit*, they entered the general culture. Davis, nutrition guru of the second half of the century explained to her avid readers that "only when protein of excellent quality is supplied can each cell function normally and keep itself in constant repair."[41] With enough high-quality protein, the body produces sufficient energy and "life is made easier." The value of a protein is related to its amino acids. Humans need twenty-two amino acids, fourteen of which can be manufactured by the body. The remaining, the so-called essential amino acids, must be obtained from food. Proteins that contain all eight are called complete or adequate. In Davis's guidelines, "protein from muscle meats, used in roasts, steaks, and chops, are complete but contain fewer of some essential amino acids that do glandular meats and are therefore less valuable. On the whole, animal proteins, such as meat, fish, eggs, milk, and cheese, contain more essential amino acids in greater abundance than do vegetable proteins; hence they have superior value.[42] A mother needed to be careful to feed her family the essential amino acids, which could most easily and assuredly be found in meat.

By the early 1970s, Davis's books sold hundreds of millions of copies, even though, or because, her claims were controversial. Davis both rejected medical authority and sought answers in the contemporary medical and scientific literature. Her spurning of physicians and their patriarchal stance, while embracing the advances of modern science and medicine, made her a heroine to millions of children of the counterculture and many middle-class mothers as well who appreciated her technical but caring approach to family nutrition. Her publications popularized the concept and sustained the ideology of meat through much of the century.

In the second half of the twentieth century, print culture denounced the saturated fat content of meat, which was linked to such health issues as hypertension, colon cancer, and diabetes. A 1984 *Newsweek* article, "America's Nutrition Revolution," addressed our "flourishing natural obsession" with eating for health. The piece noted medical evidence linking diet to heart disease and cancer and conceded that nutritional experts could not agree on what the latest evidence demonstrated as a healthful diet. Nevertheless, the authors provided rules that should govern one's diet, including lowering overall fat content and "go[ing] meatless two out of three meals." Notice: two out of three meals; meat was still the dietary requisite.[43]

Early Dietary Guidelines

We should eat meat for health; we should avoid meat for health. Though the popular press, women's magazines, and educational literature loudly argued the points, the most pervasive endorsement of a meat diet was also one of the most subtle: the federal dietary guidelines. These recommendations were and continue to be widely broadcasted in newspapers and magazines, displayed in pamphlets and brochures, and reprinted in textbooks. Guidelines refer to the whole of the diet and consequently do not seem to privilege meat. Yet, in form and substance, meat's very distinct identification confirmed it as the gold standard for protein in the human diet. Moreover, the graphics representing the guidelines significantly strengthen this emphasis.

In the late nineteenth century, food was divided into two classes: animal and vegetable. As an 1896 USDA bulletin, *Meats: Composition and Cooking*, explained: values of animal food, or meat "depend on the presence of two classes of nutrients, protein and fat. The protein is essential for the construction and maintenance of the body."[44] By the new century, however, government brochures discussed many more classes. Continuing the tradition she had initiated several years earlier in Wisconsin, Caroline Hunt and her associates at the Bureau of Home Economics used print culture to instruct women on the bases of healthful nutrition. Their bulletins, frequently reproduced by other agencies, considered milk the primary, if not exclusive, source of protein for infants; but after babyhood, they urged, other sources, such as eggs, "chicken, fish, meat, cheese, dried beans, peas, cowpeas . . . should not be overlooked. . . . [I]n order to prevent this they should be thought of as forming a *separate* and important group of food materials which are to be drawn upon more and more as the child grows."[45] This and other Department bulletins presented five essential food categories: fruits and vegetables, meats and other protein-rich foods (including milk for children), cereals and other starches, sweets, and fatty foods. "A meal, a day's ration, or a weekly food supply made up from representatives of all these five groups is likely to provide all the substances required to make the diet wholesome and attractive," government home economists assured readers.[46] They encompassed all the necessary nutrients, but they were assigned names of common foods (i.e., meat) and not nutritional components (i.e., protein). This presentation implied a nutritional equivalence among the groups and, more significantly, equated protein with meat and milk. Children obtained much of their protein from milk; adults depended on meat. The University of Wisconsin developed a scorecard to help mothers and children evaluate their meals. Its instructions specifically highlighted the role of meat in the diet of school-aged children. It advised serving an egg every morning for breakfast, prescribing "at least one serving of meat a day, if possible," and stating tolerantly that "at the other meal a meat substitute . . . may be used as the protein food."[47]

Over the years, different agencies produced different guidelines. By the 1930s, milk was often listed in a separate category and the number of categories increased. A New York State extension publication, the 1932 Cornell Bulletin for Homemakers advocated eight groups—milk, bread and cereals, vegetables, fruit and tomatoes, cod-liver oil, eggs, meat and meat substitutes, and butter and other fats—in which the significance of meat in the overall diet was apparently diminished. Despite this, the text established meat as the standard. On the one hand, it minimized meat's contribution to the family's nutrition when it advised the woman to "serve meat not more than once a day if you are watching every cent." In other words, one should consider meat only once a day *if* one needed to reduce costs; however, the homemaker could buy less expensive cuts such as shank or neck. She could also save money with other foods, but they were clearly substitutes: "Fish, when low in price, may be substituted for meat. . . . Dried peas, beans, or lentils may take the place of meat."[48] This text gave prominence to meat. In the 1910s and early 1920s, the category was most often "meat and other protein sources." Yet, within a few years, the name of the category changed dramatically. Now it was meat and meat substitutes. When substitutes are recommended, they are only substitutes, reinforcing the ideology of meat.

In the 1940s, influenced by World War II concerns about the availability of food under rationing as well as nutritional concerns, the groups were again re-designed. Some recommendations listed eight groups, others seven, still others eleven. Some combined all protein foods into one group, some separated milk into its own. Gradually these groupings added a number of daily servings, but rarely gave serving size. The Department of Agriculture's 1948 *Helping Families Plan Food Budgets* illustrated this inconsistency; it contained the Basic Seven and the Basic Eleven. The former classified foods "based on their major contribution to the nutritive content of the diet in terms of protein, vitamins, and minerals"; the latter, "for convenience in planning and flexibility" in meal planning. The Basic Seven combined meat, poultry, fish, eggs, dried beans and peas, and nuts; the Basic Eleven broke these products into three separate groups.[49] More confusing, many of the same terms were used for groups in both classifications, such as meat and milk. In another example, *Food for Little People* used ten categories to encompass its "Gold Standard for Good Eating Every Day," splitting meat, poultry, and fish, from eggs and from milk. The authors buried a casual reference to alternatives in the detailed explanation of the meat, poultry, fish group, admitting, "Dried peas and beans are fine meat substitutes and may be used once a week as such. It is inadvisable, however, to serve them to children under four. If the budget makes their use essential, cook them well; and mash them for the two- or three-year old."[50] The Department of Agriculture's 1951 pamphlet *Food for the Family with Young Children* grouped meat, poultry, fish, eggs, and dry beans or peas into one group and recommended serving at least one of these each day to each family member. However, a "Plan for the Week" dietary chart in the same publication recommended meat, poultry, or fish once daily; eggs, four or more a week per person; and dry beans and peas, nuts, one or more times a week. Even when listing alternatives, these guides were often internally contradictory or at minimum inconsistent, confusing readers while they instructed them.

The Basic Four

In the 1950s, researchers, agriculturalists, the food industry and USDA nutritionists developed new guidelines. The Basic Four that emerged from this venture—milk, meat, vegetables and fruits, and breads and cereals—shaped nutritional advice for decades. A comparison of three editions of a popular nutrition textbook demonstrated the ideology of meat and the confusion in midcentury advice.[51] In the 1949 edition of *Food for Better Living*, "A Preview of Your Food Study" presented the Basic Seven food groups. As usual, the "meat, poultry, fish, eggs, dried peas, beans" group was to be served at least once, if not twice, a day. Chapter 8, "Your Everyday Meals," reinforced this counsel: "Use of the Basic 7 Groups will make it easy for you to get all the essential nutrients in each day's meals. Whether you are eating at home, at a public eating place, or at a party remember the basic 7. Every day!" Explicitly: "Eat four or more eggs every week; eat one serving daily of meat, fish, or poultry; eat two or more servings per week of dried beans or peas, peanut butter or nuts."[52] Notice that the chart and the text differed in the number of servings per day and that the latter also slipped in additional items: peanut butter and nuts. Subsequent editions of this textbook did little to eliminate confusion. In 1960, *Food for Better Living* had increased from two to fourteen chapters. One of the new chapters, "Science Helps Us to Eat Better," also repeated the advice of one serving daily, if possible, of meat poultry, fish; four or more eggs a week, and two or more servings of dried beans, peas, nuts, including peanut butter. However, this edition introduced another method for selecting a healthful diet: a very simple chart entitled "Food for Fitness—the Basic Four: A Daily Food Guide" with only four groups—milk, meat group, vegetable-fruit, and bread-cereal—and the explanation: "In using this Daily Food Guide you select the main part of your diet from four broad food groups. To this you add other foods to make your meals more appealing and satisfying." The meat group contained foods "valued for their protein, which is needed for growth and repair of body tissues."[53] In the chapter on protein foods, eleven pages are devoted to recipes for eggs, five for those with dried beans and peas, and fifty-one for meats—and the name of the group is meat, not protein: both the recipes and the name reinforced the significance of meat in the daily diet.

In the 1960s, educators recognized the confusion inherent in so many different schemas. Indeed, the Basic Four Guide had been "developed to simplify as much as possible the daily selection of food intake with the belief that people will use a simplified plan more effectively and willingly than a more complicated one."[54] Consequently, in the next edition of *Food for Better Living*, now titled *Food for Modern Living* (1967), the authors dropped any mention of the Basic Seven food groups. They discussed only the Basic Four food guide, including the "meat group" which "supplies *complete* proteins," with the recommendation of two or more servings daily. The section on recipes for the "meat group" still included sixteen pages on meat alternatives, such as eggs, dried beans and peas, and peanut butter, but the 1967 edition devoted sixty-four pages to meat, poultry, and fish.[55] Again, the name of the group was "meat."

With the Basic Four, meat gained in prominence, now one in four rather than one in seven. The meat industry was clearly aware of the power of print culture to inform our nation's understanding of nutrition and the significance of food groups. However, industry leaders objected to the addition of portion size in the guidelines; the recommended two to three ounces of cooked meat was considerably smaller than the typical contemporary portion size (and significantly smaller than today's typical serving size). Milk, meat, vegetables and fruits, and breads and cereals were given relatively equal weight in the diet. The recommendations within the meat category included beans and peas as alternative protein sources, but these appeared more as an aside. The title was "meat," not "protein sources." Diet lists published in widely circulated magazines such as *Parents* and *Good Housekeeping* counseled mothers to provide for two or more servings of meat, poultry, fish, or eggs every day, tolerating "beans, peas, and nuts as an occasional substitute."[56] Many nutrition advisors simply recommended "good quality protein foods," such as "meat, milk, poultry and fish."[57] Whether or not the advice specified alternatives, what is most critical is that the name of the group was "meat." Millions of schoolchildren learned the basic four groups as milk, meat, fruits and vegetables, and breads and cereal, the foods needed for a well-balanced, healthful diet. (An informal survey of people who would have been students in primary or secondary schools in this period documents the lingering influence of this education; most remember learning about the Basic Four. Though they could not necessarily list all four, they invariably commented on meat.) Meat was a vital component of the family's nutrition.

An example of this dietary advice is the government's 1970 pamphlet *Family Fare: A Guide to Good Nutrition*, which instructed mothers to provide "some meat, poultry, fish, eggs, milk or cheese at each meal." The meat category is very broad, encompassing beef, veal, lamb, pork, variety meats (such as liver, heart, kidney), poultry and eggs, fish and shellfish. *Family Fare* mentioned, but did not pay much attention to, alternatives to animal protein such as dry beans, dry peas, lentils, nuts, peanuts, and peanut butter. To be sure, "proteins from legumes, especially soybeans, chickpeas, and peanuts are almost as good as proteins from animal sources," but almost as good is not as good. Consequently, the mother was reminded, "combining cereal and vegetable foods with a little meat or other source of animal protein will improve the protein value of the meal."[58] Clearly, the protein standard is meat.[59]

In the late 1970s, the U.S. government produced a series of leaflets for elementary school students, including *Mary Mutton and the Meat Group*. Mary asked "why is meat good for people."[60] Heidi Heifer explained that meat is rich in protein; Leonard Lamb explained why protein is important in our diet. To help children understand the variety of meats available for good nutrition, Polly Pig constructed three crossword puzzles: for beef, pork, and lamb. Mary Mutton reminded the children about poultry, after which Heidi Heifer and Gussie Goose mentioned "fish and eggs, and even dried peas and beans get counted as part of the meat group."[61] Note again, the title is meat group. Alternate sources of protein are mentioned in passing only. Meat remains paramount.

Concerns about Meat
in the Late Twentieth Century

In 1960s and the 1970s, an increasingly vehement discussion of the role of dietary fat in chronic disease challenged the ideology of meat.[62] Many writers recommended that we decrease our intake of fat, that is, eat less of those foods containing fat and cholesterol, most especially red meat. The *Dietary Goals for the United States*, released January 1977, advised the American consumer to "reduce consumption of meat." Once again, recognizing the authority of print culture in shaping nutritional practices, the meat industry protested what leaders saw as a denigration of a significant food source. They pressured the government to revise the goals that appeared to slight meat. The new report, issued later that year, read "choose meats, poultry, and fish which will reduce saturated fat intake."[63] The goals replaced the straightforward message—"reduce consumption"—with a complex statement demanding that consumers learn all about and then find the meats, and poultry, and fish with the desired characteristics. Even the more simplified 2005 version continues to privilege meat: "Go lean with protein. Choose lean meats and poultry. Bake it, broil it, or grill it. And vary your protein choices—with more fish, beans, peas, nuts, and seeds."[64] But notice again, "meats and poultry," with "fish, beans, peas, nuts, and seeds" as an afterthought.

The story behind the food pyramid, the graphic that replaced the Basic Four Food Groups in the early 1990s, demonstrates the power the meat industry accorded print culture. The USDA's Human Nutrition Information Service (HNIS) wanted to illustrate dietary advice in a form referring to nutrients such as fat, salt, and sugar rather than foods that contained them. The agency contracted with a consumer research firm to develop a design emphasizing the importance of eating a variety of foods in appropriate portions, while moderating ingredients such as fat and sugar. Nutrition experts reviewed the proposal, which was also presented in professional meetings and discussed in the public media. The HNIS anticipated publication of the final food pyramid in March 1991, but it was summarily withdrawn before that because of objections from the cattle industry, the sugar industry, nutritionists and other scientists, and various politicians and government bureaucrats. The meat partisans were particularly incensed, fearing that the proposed representation would bias readers against meat. Members of the National Cattlemen's Association board complained to Agriculture Secretary Edward R. Madigan on April 15. In an April 19 letter to the secretary, the Meat Institute added its complaint.[65] In *Food Politics: How the Food Industry Influences Nutrition and Health*, Marion Nestle details the political machinations of industry lobbying; it took another year and $855,000 more before the revised food pyramid was released. Nestle has identified thirty differences between the original and the revision, but only two are significant here. First, the name "Eating Right" was dropped; "Eating Right" was the title of a series of prepared foods sold by Kraft Foods, and rivals feared that with that name the food pyramid gave Kraft an advantage in the marketplace. Second, the number of servings per day was moved outside the graphic and set in boldface type, to suggest *at least* two to three servings. Moreover, in the accompanying text, portion size was increased. True, the category now listed not just "meat" but also poultry, fish, dried beans, eggs, and nuts. However, that the meat industry fought so hard for these changes indicates its belief in the power of the graphic and of print culture to keep meat foremost in the American diet.

Debates over the food pyramid have continued. Nutritionists and scientists in particular insist that health is more than what we eat and that the same advice is not applicable to all. Consequently, twelve different food pyramids were developed. Intended to better represent nutritional advice and reflect specific ethnic and therapeutic diets, they have instead created extremely confusing graphics that have not yet proven popular.[66] Notably, the text accompanying the food pyramids continues to give precedence to meat. For example, in 1994 the Food and Nutrition Service of the USDA brochure *What's in a Meal?* explained to day care personnel and others which foods are reimbursable under the Child and Adult Care Food Program. Its food pyramid had one section marked "Meat, poultry, fish, dry beans, eggs, & nuts group," but the text used the phrase "meat or meat alternate"; once again, meat is the standard and anything else merely an "alternate."[67]

More recently, two somewhat different challenges to the ideology of meat have arisen out of concerns for the safety of the American population and the health of the planet. Previously, arguments against meat had primarily centered on the health of the individual. Now, factory farms in which cattle are housed in close quarters and fed high-protein grains have increased the risk of producing and distributing contaminated products.[68] High-profile media stories have educated the public in some of the newest bacteriological and biological discoveries, using science to alert consumers to the potential risk of eating infected and drugged products. Over the years, American consumers have received crash courses in the action of hormones in cattle, the role of antibiotics in food production, and the consequences of ingesting E. coli [0157:H7], salmonella, listeria, campylobacter, bovine spongiform encepalopathy (BSE, or mad cow disease), and the like. However, fear of contamination and adulteration has had little impact on our meat consumption. For instance, in December 2003, the government announced that BSE had been identified in a Washington State cow. Consumer surveys forecast that beef demand could fall by as much as 15 percent. Surprisingly, one study of household beef purchases between 1998 and 2004 noted: "The magnitude of responses in the market was difficult to estimate precisely, but the duration was clear; within two weeks, consumers were behaving exactly as they had before the announcement."[69] Over longer periods, the consumption of red meat has decreased, for example, between 1970 and 1999, it dropped by 11 percent. But in the same period, poultry consumption increased 102 percent. So our love affair with red meat may be cooling slightly, but not with meat per se.[70]

The demand for meat products that stimulated the spread of factory farms also threatened the health of the environment. These firms consume enormous amounts of energy, pollute nearby water supplies, and generate about one fifth of the world's greenhouse gasses.[71] Moreover, they squander grains. United Nations studies estimate that approximately 800 million on the planet suffer hunger or malnutrition, but we feed the majority of our corn and soy to farm animals. To produce these grains, agriculture has spread into new regions of the world, most especially devastating rain forests in countries like Brazil. At the same time, grain prices have nearly doubled. Using farm animals to convert grains to calories is shortsighted and costly. It takes between two and five times more grain to produce the same number of calories from farm animals as from direct consumption of the grain. Put another way, one needs up to forty pounds of grain to produce one pound of beef.[72]

Despite these challenges, meat remains the primary component of the U.S. diet. In some regards, food is equated with meat. The next time you pick up a prescription from the pharmacist with instructions to take your medicine with food, look carefully at the bottle (see figure 16). Study the icon used to remind you of these directions: it could be a veggie burger, but it looks distinctly like a hamburger. Such an image promotes meat not simply as a good source of protein, but as *the* definition of food.

Figure 16. Label for prescriptions to be taken with food. (Courtesy of Neuhauser Pharmacy, Madison, Wisconsin)

Conclusion

The Basic Four, the Food Pyramid, and Molly Mutton, elements of print culture disseminated by the federal government, clearly and subtly bolster the ideology of meat. At the same time, print culture has bombarded the American public with scientific arguments against meat. Early in the century, it was concern for individual health, for over-consumption of protein; later, for fat. By the post-war era, scientifically derived evidence of contamination magnified worries about the welfare of the population. In each of these cases, scientists have taken their evidence to the public and have argued their points in the media. So far, neither has had any lasting impact on our consumption patterns. Will their arguments have more effect today? Will more global issues, such as the need to feed the world and save the rain forests, change American eating habits? Apparently, scientific arguments alone are not sufficient. If we are serious about promoting a different sort of diet, one in which meat is not the gold standard, we must be aware of the power of not only scientific claims but also of their presentation and the influence of print culture. We need to invent a new schema for nutrition education, textual and graphic, that does not overtly or inadvertently reinforce the ideology of meat.[73]

Notes

1. For more on the relationship between science and medicine and motherhood in the United States, see Rima D. Apple, *Perfect Motherhood: Science and Childrearing in America* (New Brunswick, N.J.: Rutgers University Press, 2006).

2. W. O. Atwater, *Methods and Results of Investigations on the Chemistry and Economy of Food*, U.S. Office of Experiment Stations Bulletin, no. 21 (Washington, D.C.: USDA, 1895).

3. Ellen Alden Huntington, *The Fireless Cooker*, Bulletin of the University of Wisconsin, no. 217 (Madison, Wisc., 1908), available in University of Wisconsin–Madison Archives, ser. 10/4/1, box 1.

4. Rima D. Apple, *The Challenge of Constantly Changing Times: From Home Economics to Human Ecology at the University of Wisconsin–Madison, 1903–2003* (Madison: Board of Regents of the University of Wisconsin System, 2003), esp. 1–14; Marjorie East, *Caroline Hunt: Philosopher for Home Economics* (University Park: Pennsylvania State University, 1982).

5. Mrs. S. T. Rorer, "Why Cooking Is So Easy For . . ." *Ladies' Home Journal*, September 1907, 46.

6. J. B. Harrington, "Teaching Cooking to Prevent Crime," *Forecast* 11 (1916): 57–58.

7. Cornell Bulletin of Homemakers, part 1 (1903), 7. See also Caroline Benedict Burrell, "Mother's Responsibility in Regard to Diet," in *The Child Welfare Manual: A Handbook of Child Nature and Nurture for Parents and Teachers* (New York: University Society, 1916), 218–20.

8. See, for example, Burrell, "Mother's Responsibility in Regard to Diet."

9. Josephine H. Kenyon, "How to Feed Your Children—in Wartime," *Good Housekeeping*, May, 1942, 68, 192. See also *Meat for Thrifty Meals*, Home and Garden Bulletin, no. 27 (Washington, D.C.: USDA, Bureau of Human Nutrition and Home Economics, 1953); "Summer Meals to Suit Your Pocketbook," *Good Housekeeping*, January 1935, 92; *A Week's Food for an Average Family*, Farmers' Bulletin, no. 1228 (Washington, D.C.: USDA, 1921).

10. See, for example, *Nutrition Education in the Elementary School*, Nutrition Education series, pamphlet no. 1 (Washington, D.C.: Federal Security Agency, U.S. Office of Education, 1943).

11. Lola T. Dudgeon, *Food Makes a Difference*, Cornell Extension Bulletin, no. 775 (Ithaca, N.Y.: Cornell University, November 1949). See also Mary Hinman Abel, *Successful Family Life on the Moderate Income* (Philadelphia: J. B. Lippincott Co., 1921); "Child Welfare and Economics," *American Food Journal* 14 (December 1919): 8; "Improving Your Family's Health," *Parents*, October 1941, 54, 59, 61; Louella G. Shouer, "Is Your Family Well Fed?" *Ladies' Home Journal*, December 1941,114–15.

12. On women and nutrition, Rima D. Apple, "Science Gendered: Nutrition in the United States, 1840–1940," in *The Science and Culture of Nutrition, 1840–1940*, ed. Harmke Kamminga and Andrew Cunningham (Amsterdam: Rodopi, 1995), 129–54. For the continuing primacy of women's responsibility, see Katherine S. Tippet and Linda E. Cleveland, *Results from USDA's 1994–1996 Diet and Health Knowledge Survey*, NSF Report, no. 96-4 (Washington, D.C.: USDA, Agricultural Research Service, May 2001), table 11.

13. "Brief Discussion of Human Nutrition," *Cornell Reading-Course for Farmers' Wives*, series 3, no. 13 (Ithaca: New York State College of Agriculture, 1905).

14. Emma E. Pirie, *The Science of Home Making: A Textbook for Home Economics* (Chicago: Scott, Foresman and Co., [1915]).

15. Jane Brody, "America Leans to a Healthier Diet," *New York Times*, 13 October 1985.

16. Judy Putnam and Shirley Gerrior, "Trends in U.S. Food Supply, 1970–1997," in *America's Eating Habits: Changes and Consequences*, ed. Elizabeth Frazao, Agriculture Information Bulletin, no. 750 (Washington, D.C.: U.S. Department of Agriculture

Economics Research Service, 1999), 136 (http://www.ers.usda.gov/Publications/ AIB750/).

17. Philip H. Dougherty, "Advertising: Dressing Pork for Success," *New York Times*, January 15, 1987.

18. Anne Pierce, "Meat in the Child's Dietary," *Parents*, April 1930, 30; Jackie Plant and Fraya Berg, "Healthy Eating, Happy Kids," *Parents*, January 2007, 103–4, 106, 109.

19. Ida Cogswell Bailey-Allen, "The Every-Day Chemistry of Foods and Cookery," *Good Housekeeping*, July 1915, 128. See also Harvey Levenstein, *Revolution at the Table: The Transformation of the American Diet* (New York: Oxford University Press, 1988), 23–24; Susan Williams, *Savory Suppers and Fashionable Feasts: Dining in Victorian America* (Knoxville: University of Tennessee Press, 1996), esp. 98–102; Waverly Root and Richard de Rochemont, *Eating in America: A History* (Hopewell, N.J.: Ecco, 1976), esp. 189–212; Elaine N. McIntosh, *American Food Habits in Historical Perspective* (Westport, Conn.: Praeger, 1995).

20. Mary Swartz Rose, *Everyday Foods in War Time* (New York: Macmillan Co., 1918), 23.

21. "What the American Eats," *American Food Journal* 15 (February 1920): 37–38.

22. "Our Diet Changing," *American Food Journal* 15 (March 1920): 7.

23. Karen Iacobbo and Michael Iacobbo, *Vegetarian America: A History* (Westport, Conn.: Praeger, 2004), 158.

24. Those with household incomes of $1,000 to $2,000 spent 29.4 percent; of $3,000 to $4,000, 28.9 percent; of $5,000 to $7,500, 30.7 percent. The amount purchased varied by income: those in the $1,000 to $2,000 group bought 8.7 pounds during one week in the spring, those in the $5,000 to $7,500 bought 11.3 pounds. Faith Clark et al., *Food Consumption of Urban Families in the United States . . . With an Appraisal of Method of Analysis*, Agriculture Information Bulletin, no. 132 (Washington, D.C.: USDA, Home Economics Branch, Agriculture Research Service, October 1954), 2–3.

25. Ercel Eppright, Mattie Pattison, and Helen Barbour, *Teaching Nutrition*, 2nd ed. (Ames: Iowa State University Press, 1963), 37. See chapter 3, "What We Eat," 36–66.

26. *Food Consumption: Households in the United States, Season and Year 1977–1978*, NFCS 1977–78, report no. H-6 (Washington, D.C.: USDA, Human Nutrition Information Service, Consumer Nutrition Division, June 1983), 1–3.

27. Tippet and Cleveland, *Results from USDA's 1994–1996 Diet and Health Knowledge Survey*, 5.5a, 5.5b, 5.4a, 5.4b, 5.3a, 5.3b. The group described was "meat, poultry, fish, dry beans, eggs, and nuts," but it is clear from other reports that the last three listed did not contribute significantly to total consumption. See, for example, Helen Smiciklas-Wright et al., *Foods Commonly Eaten in the United States: Quantities Consumed per Eating Occasion and in a Day. 1994–1996* (Washington, D.C.: USDA, Agriculture Research Service, January 2002), appendix table B.

28. Iacobbo and Iacobbo, *Vegetarian America* 131–32.

29. "What about Meat?" *Hygeia* 17 (1939):1062–63, and "Meat and the Authority Behind It . . . ," advertisement, *Journal of Home Economics* 33, no. 6 (June 1941): 19.

30. Hazel Kepler and Elizabeth Hessler, *Food for Little People* (New York: Funk & Wagnalls, 1950), 21.

31. Pat Baird, "The Diet Women Need Most," *Redbook*, April 1992, 123–26; Daryn Eller, "Should You Eat Meat?" *Redbook*, March 1993, 130, 134; Carol Krucoff, "Exercise

and Iron," *Saturday Evening Post*, September-October 1996, 21; Jackie Storm, "Did Mother Know Best?" *Women's Sports and Fitness*, January-February 1991, 18–26.

32. See, for example, Whitman H. Jordan, *Principles of Human Nutrition: A Study in Practical Dietetics* (New York: Macmillan Co., 1912); Elizabeth Robinson Scovil, "The Child of Three and Over," *Ladies' Home Journal*, November 1902, 47. Vegetarianism could mean the avoidance of all animal products, the shunning of animal flesh only, or something in between. Today, surveyors do not ask, "Are you a vegetarian?" Instead, they ask which of the following foods, if any, you never eat—then they list meat, poultry, fish or seafood, dairy products, eggs, and honey. Charles Stahler, "How Many Adults Are Vegetarian? The Vegetarian Resource Group Asked in a National Poll," *Vegetarian Journal* 25, no. 4 (2006) (www.vrg.org/jounral/vj2006issue4/vj2006issue4poll.htm).

33. John C. Olsen, *Pure Foods: Their Adulteration, Nutritive Value, and Cost* (Boston: Ginn and Co., 1911), 78–80.

34. Mary Davis Swartz, "How Much Meat?" *Good Housekeeping*, January 1910, 106–9. Other examples include George F. Butler, "Why So Much Meat? One Important Aspect of the Big Problem of the High Cost of Living," *Ladies' Home Journal*, October 1912, 30; Carlotta C. Greer, *School and Home Cooking* (Boston: Allyn & Bacon, 1920); "Meat—Its Uses and Abuses," *Delineator* 67 (February 1906): 360–62; S. T. Rorer, "Why I Do Not Believe in Much Meat," *Ladies' Home Journal*, April 1905, 39.

35. Rose, *Everyday Foods in War Time*, 31.

36. Bailey-Allen, "Every-Day Chemistry of Foods and Cookery," 128.

37. F. J. Schlink, *Eat, Drink, and Be Wary* (New York: Covivi Friede, 1935), 81–86.

38. "How Much Food Do We Need," *American Food Journal* 14(July 1919): 15–16. For more on vitamins, see Rima D. Apple, *Vitamania: Vitamins in American Culture* (New Brunswick, N.J.: Rutgers University Press, 1996).

39. For examples, see "How Much Food Do We Need" and Butler, "Why So Much Meat?"

40. Anne Pierce, "Meat Cookery," *Parents*, November 1930, 44.

41. Adelle Davis, *Let's Eat Right to Keep Fit* (1954; repr., New York: New American Library, 1970), 29. For earlier discussions, see, for example, B. W. Gardner, "Value of Meat in the Diet," *Consumers' Research Bulletin*, no. 11 (January 1943): 15–17; C. Robert Moulton, "The Value of Meat in the Diet," *American Food Journal* 18 (December 1923): 559–61, 90; Elsie Fjelstad Radder, "The Growing Child Eats Meat," *Hygeia* 6 (June 1928): 343–44; Paul Rudnick, "The Nutritive Values of Meats," *American Food Journal* 16 (September 1921): 12–13. Also Frances R. Godshall, *Nutrition in the Elementary School*, ed. Helen R. LeBaron, Harper's Home Economics series (New York: Harper & Brothers, 1958), 17–19.

42. Davis, *Let's Eat Right to Keep Fit*, 30, 34–35. Vegetarians of this period accepted the need for complete proteins in the diet and constructed complicated recipes and meal plans that ensured the consumption of all necessary amino acids at the same meal without the inclusion of meat. See, for example, Ellen Buchman Ewald, *Recipes for a Small Planet: The Art and Science of High Protein Vegetarian Cookery* (New York: Ballantine, 1973); Frances Moore Lappe, *Diet for a Small Planet* (New York: A Friends of the Earth/Ballantine Book, 1971).

43. Jean Seligmann et al., "America's Nutrition Revolution," *Newsweek*, 19 November 1984, 111.

44. Charles D. Woods, *Meats: Composition and Cooking*, Farmers' Bulletin, no. 34 (Washington, D.C.: USDA, 1896), 5.

45. Caroline L. Hunt and Mabel Ward, *School Lunches*, Farmers' Bulletin, no. 712 (Washington, D.C.: USDA, 1916), 8 (emphasis added).

46. *A Week's Food for an Average Family*, 4. See also Caroline L. Hunt, "Food for Young Children" (Washington, D.C.: USDA, 1916); Caroline L. Hunt, "Good Proportions in the Diet" (Washington, D.C.: USDA, 1923); Caroline L. Hunt and Mabel Ward, "School Lunches," *Red Cross Course in Food Selection* (Washington, D.C.: American Red Cross, 1921).

47. Helen Parsons, *Score Cards for Children*, Special Circular (Madison: Extension Service of the College of Agriculture, University of Wisconsin, 1924), 5.

48. Mary Henry and Day Monroe, *Low Cost Food for Health*, Cornell Bulletin for Homemakers, no. 236 (Ithaca: New York State College of Home Economics at Cornell University, 1932), 3, 6.

49. *Helping Families Plan Food Budgets*, Miscellaneous Publication, no. 662 (Washington, D.C.: USDA, 1948).

50. Kepler and Hessler, *Food for Little People*, 14–15, 21.

51. Irene E. McDermott, Mabel B. Trilling, and Florence Williams Nicholas, *Food for Better Living* (Chicago: J. B. Lippincott Co., 1949); Irene E. McDermott, Mabel B. Trilling, and Florence Williams Nicholas, *Food for Better Living* (Chicago: J. B. Lippincott Co., 1960). Irene E. McDermott, Mabel B. Trilling, and Florence Williams Nicholas, *Food for Modern Living* (Philadelphia: J. B. Lippincott Co., 1967).

52. McDermott, Trilling, and Nicholas, *Food for Better Living* (1949), 9, 367, 369.

53. McDermott, Trilling, and Nicholas, *Food for Better Living* (1960), 31, 32, 32b.

54. Eppright, Pattison, and Barbour, *Teaching Nutrition*, 107–8.

55. McDermott, Trilling, and Nicholas, *Food for Modern Living*, 175, 376–455.

56. "The Better Way: Is Your Family Eating Sensibly?" *Good Housekeeping*, September 1968, 169–71; "Everyday Nutrition," *Parents*, April 1957, 97; Ginny McCarthy, "What It Takes to Feed Your Family Well," *Parents*, November 1966, 94–95, 152.

57. Charles Glen King, "Nutrition Research Is Vital to Family Health," *Parents*, October 1961, 100–103, 49–50, 52.

58. Agricultural Research Service Consumer and Food Economics Research Division, *Family Fare: A Guide to Good Nutrition*, Home and Garden Bulletin, no. 1 (Washington, D.C.: USDA, 1970), 7, 8.

59. C. Brock, "The ABC's of Family Meal Planning," *Parents*, February 1970, 72–73; Better Homes and Gardens, *Nutrition for Your Family* (Des Moines, Iowa: Meredith Corp., 1978).

60. *Fred, the Horse Who Likes Bread* ([Washington, D.C.]: USDA, Office of Governmental and Public Affairs, 1979); *Gussie Goose Introduces the Fruit and Vegetable Group* ([Washington, D.C.]: USDA, Office of Governmental and Public Affairs, 1979); *Mary Mutton and the Meat Group* ([1979]); USDA, Office of Governmental and Public Affairs, *Meet Molly-Moo: She's Blue, but You Can Make Her Happy, Here's What You Can Do* ([Washington, D.C.]: USDA, Office of Governmental and Public Affairs, 1979).

61. USDA, *Mary Mutton and the Meat Group*, 3.

62. For an example of this as presented in the popular press, see "The Better Way."

63. Marion Nestle, *Food Politics: How the Food Industry Influences Nutrition and Health* (Berkeley: University of California Press, 2002), 40–42.

64. "Finding Your Way to a Healthier You: Based on the Dietary Guidelines for Americans" ([Washington, D.C.]: U.S. Department of Health and Human Services, U.S. Department of Agriculture, 2005) (www.health.gov/dietaryguidelines/dga2005/document/pdf/brochure.pdf).

65. Marian Burros, "Are Cattlemen Now Guarding the Health House?" *New York Times*, May 8, 1991.

66. On creating your own food pyramid, see Carol Ann Rinzler, *Nutrition for Dummies* (Foster City, Calif.: IDG Books Worldwide, 1999), 207–8.

67. *What's in a Meal? Resource Manual for Providing Nutritious Meals in the Child and Adult Care Food Program* (Chicago: USDA, Food and Nutrition Service, [1994]).

68. For example, Eric Schlosser, "Bad Meat: The Scandal of Our Food Safety System," *Nation*, September 16, 2002, 6–7; Eric Schlosser, "Order the Fish," *Vanity Fair*, November 2004, 240, 43–46, 55–57; Micah L. Sifry, "Food Money," *Nation*, December 27, 1999, 20.

69. Fred Kuchler and Abebayehu Tegene, "Did BSE Announcements Reduce Beef Purchases?" (Washington, D.C.: USDA, Economic Research Service, 2006), iii (www.ers.usda.gov/publicatin/err34/err34.pdf). See also Brain Coffey et al., *The Economic Impact of BSE on the U.S. Beef Industry: Product Value Losses, Regulatory Costs, and Consumer Reactions* (Kansas State University, April 2005); Fred Kuchler and Abebayehu Tegene, "U.S. Consumers Had Short-Term Response to First BSE Announcements," *Amber Waves* 5, no. 4 (2007): 24–9.

70. Arbindra P. Rimal, "Factors Affecting Meat Preference among American Consumers," *Family Economics and Nutrition Review* 14, no. 2 (2002): 36–43.

71. Mike Kiesius, "Ready to Go Green," *AARP Bulletin*, May 2008, 43. To put this in perspective, one estimate suggests that if we ate 25 percent less meat, we would reduce greenhouse gasses by 1.9 million tons.

72. Terry J. Allen, "As Hunger Rises, Chew on This," *In These Times*, May 2008, 45. Groups such as People for the Ethical Treatment of Animals echo age-old ethical arguments for vegetarianism; it is, however, a minor theme at the turn of the twenty-first century.

73. Since this chapter was presented, the United States government has replaced the food pyramid with the food plate (http://www.choosemyplate.gov/index.html, accessed February 21, 2012). This graphic includes five groups: fruits, vegetables, grains, protein, and dairy. The protein group contains beans and peas, processed soy products, and nuts and seeds, especially listed for vegetarians. It is too soon tell how effective this plate will be in altering our diets.

What Two Books Can (and Cannot) Do

Stewart Udall's The Quiet Crisis and Its Twenty-Fifth Anniversary Edition

CHERYL KNOTT

Priscilla Coit Murphy, in *What a Book Can Do: The Publication and Reception of Silent Spring*, writes: "What a book *does* . . . can tell us what we use books for, giving us the opportunity to consider . . . whether any other medium would perform the same function or fill the same needs better than, or even as well as, a book." Murphy provides ample evidence that a book, especially a heavily documented nonfiction work of advocacy such as Rachel Carson's *Silent Spring*, can galvanize public opinion, even when it is not read by a lot of the people who take action because of it. We could assume that the immediacy of new media such as blogs and instant messaging might have obviated the need for books, particularly for moving people to action regarding a timely political or environmental issue. But Murphy asserts, "Authors in every field with every purpose, even journalists and public figures who already have direct access to the media, have continued to resort to books to communicate their messages to the public at large. Moreover, book publishing, far from collapsing with the advent of each new 'competitor' medium, has continued to survive and even to flourish."[1] According to Murphy, it is the book (as opposed to the author) that launches social and political movements as it takes on a life of its own in ways the author and publisher could not have foreseen.

Silent Spring remains in print. Although new editions with forewords and afterwords by different authors have appeared, the book's core text has never been revised, because Carson died soon after it was first published. Comparing later instances of Carson's own revisions in response to the first edition's popular reception and the political, industrial, and practical impact of Carson's work could have helped round out a discussion of "what we use books for" as those uses adjusted to changing contexts.

A book published a year after Carson's offers an opportunity to begin to do that. In 1963, Stewart Lee Udall, secretary of the interior under Presidents Kennedy and Johnson, published *The Quiet Crisis*, discussing the degradation of the nation's natural resources and providing a history of the American conservation movement. It became a *New York Times* best seller, widely reviewed in the popular press and in scientific journals. In 1988, Udall, no longer a public official, issued an updated twenty-fifth anniversary edition with additional chapters, including one acknowledging Carson's contribution to the environmental movement. The book did not receive the same notice as the first edition. The Udall papers at the University of Arizona along with published reviews and notices make it possible to evaluate the publication and reception of the two versions of *The Quiet Crisis* in an attempt to understand their impact in the differing contexts of environmental science and politics in 1963 and 1988.

Stewart Lee Udall

Stewart Udall was born in 1920 in St. John's, Arizona, the town his Mormon grandfather had moved to as a missionary, a few miles from the state's border with New Mexico. He earned a law degree at the University of Arizona and practiced law in Tucson from 1948 to 1954, when he made a successful run for the U.S. House of Representatives as a Democrat and was elected to three terms. After ensuring that Democrats in the state of Arizona would go for John F. Kennedy as their presidential nominee, in 1961 he became the newly elected President Kennedy's secretary of the interior, from which post he crafted an ambitious environmental protection program as part of Kennedy's New Frontier. Although Udall had not supported Lyndon Johnson's bid for the White House in 1960, Johnson kept him on for the duration of his administration. Udall and Lady Bird Johnson collaborated on environmental efforts such as beautification, and each had an influence on President Johnson's policies to protect the environment. Remaining secretary until 1969, Udall oversaw about fifty thousand employees who spent some $800 million annually to discharge the Department of Interior's duties involving the management of 750 million acres of federal land, including the national park system; the management of natural resources; cartography, particularly involving geological and topographical features; and land reclamation, among other things.[2]

According to a historical account of the Interior Department, Udall expanded the department's focus beyond the raw materials and natural resources of the West and worked with Congress to pass the "Federal Clean Air Act of 1963, the Wilderness Act of 1964, the Land and Water Conservation Fund Act of 1965, the National Historic Preservation Act of 1966, the Wild and Scenic Rivers Act of 1968, and amendments strengthening the Federal Water Pollution Control Act of 1956."[3]

The Context of Crisis

Udall had a lifelong love of the outdoors and activities such as hiking and camping. He revered President Theodore Roosevelt and his chief forester, Gifford Pinchot, and President Franklin Delano Roosevelt and his interior secretary, Harold Ickes, for their dedication to preserving natural resources. Pinchot and Ickes served as his models for activist cabinet members involved in policy formulation.[4]

President Theodore Roosevelt had held a conference on conservation during his administration fifty years before Kennedy took office, and Secretary Udall similarly arranged for a White House Conference on Conservation to be held in May 1962. Since 1958, Rachel Carson had been researching and writing what would become her iconic work of environmental activism, *Silent Spring*. Her interviews and correspondence regarding government research and regulation of pesticides had alerted individuals in executive agencies such as the Department of the Interior's Fish and Wildlife Service (where Carson had worked previously as a biologist), Congress, and the judiciary, as well as in the chemical industry, and chapter drafts were already circulating among scientists who were checking her facts, interpretations, and tone. Udall invited Carson to the conference, and he subsequently "assigned a member of his staff to track the book's reception and report ideas for future policy initiatives."[5]

In addition to hosting the White House conference, Udall attended meetings, made speeches, escorted the president on trips to scenic western landmarks, applied pressure on various key U.S. congressional representatives and senators, and planned and strategized his conservation agenda. He also turned to a tool whose potential for high impact Rachel Carson was demonstrating: he wrote a book. It was his first. With *The Quiet Crisis*, Udall launched a publishing career that would include *1976, Agenda for Tomorrow* (1968); *America's Natural Treasures: National Nature Monuments and Seashores* (1987); *To the Inland Empire: Coronado and Our Spanish Legacy* (1987); *The Quiet Crisis and the Next Generation* (1988); *The Myths of August, A Personal Exploration of Our Tragic Cold War Affair with the Atom* (1994); and a reissue of *The Inland Empire* titled *Majestic Journey: Coronado's Inland Empire* (1995).

Other politicians, particularly presidential candidates, had established a tradition of issuing books as statements of personal and political philosophy.

John Kennedy's *Profiles in Courage*, which highlighted instances in which U.S. senators were willing to take stands on difficult issues and written while Kennedy was himself a senator, won a Pulitzer Prize in 1957 and broadened the author's national reputation just in time for his presidential run. Although executive agency heads and members of the president's cabinet had published books—most often memoirs issued after they left office, such as Gifford Pinchot's *The Fight for Conservation* (1910) and *Breaking New Ground* (1947); *The Reminiscences of Carl Schurz* (1908); and *The Memoirs of Cordell Hull* (1948)—it was not common among sitting executive department heads. An exception was Harold Ickes, who published his *Autobiography of a Curmudgeon* (1943) a decade into his thirteen-year term as secretary of the interior. More typically, officials in the federal executive branch published narrowly focused reports and addresses on specific issues representing the agency's activities. Issued as government documents, such texts were in the public domain. In contrast, Udall's books, issued by commercial publishers, were copyrighted in his name.

The Quiet Crisis is related to what Jeffrey Tulis has named the "rhetorical presidency," characterized by the chief executive's attempts to persuade the people directly rather than, or in addition to, working through their elected representatives in Congress. Tulis analyzes presidential speeches and addresses as evidence of selected presidents' relative success at interpreting and reinterpreting their constitutional powers and roles.[6] Other scholars of political communication similarly mine presidential speeches and addresses, including printed texts and transcripts of broadcasts, to understand the executive's attempts to shape and effect policy.[7] Political scientists have paid somewhat less attention to the book as an indicator of policy and power formulation. Yet the book offers some advantages—for the writer and the reader—that speeches and shorter texts do not. As Michiko Kakutani asserts in a *New York Times* review of books by eight presidential hopefuls:

> Most books by politicians are, at bottom, acts of salesmanship: efforts to persuade, beguile or impress the reader, efforts to rationalize past misdeeds and inoculate the author against future accusations. And yet beneath the sales pitch are clues—in the author's voice, use of language, stylistics and self-presentation—that provide some genuine glimpses of the personalities behind the public personas. In short, when candidates decide to publish, they can still run, but they can't hide—at least not entirely.
>
> At the same time these candidates' books remind us that the ability to construct a powerful narrative is an essential skill for a politician, for it confers the ability to articulate a coherent vision of the world, to make sense of history and to define the author—before he or she is defined by opponents and the news media.[8]

Although Udall was not president and apparently not planning to run for president and thus was not engaged in sharing his autobiography with voters, he was in a powerful position to design and implement domestic environmental conservation and protection policies. A book by the interior secretary offered the same key advantage that a presidential candidate's book did: control over the message (or at least its expression). A politician's book is the one place where he can deliberate over the choice of words, what to put in, what to leave out, and how to finesse the difficult things (in the autobiographical genre, marijuana smoking, adultery) that must be aired. The book is a play for credibility and likability, an account of one's thinking and an acknowledgement of the influences on that thinking, an assertion of one's values and beliefs, and a statement of intentions. It is the place where the politician's communication can be most nuanced. The book gives the author control over how he expresses himself, even though it cannot give him control over how the information is received and used. A politician's book represents the author's best attempt to communicate clearly the information he wants to convey to his intended audience. And even when the book is only one channel among the many media outlets exploited in the quest for convincing communication, it remains a constant the politician can point to during subsequent speeches and interviews. For presidents such as Jimmy Carter, who continue to publish books long after their terms of office have ended, this last function is perhaps the most important. Carter's efforts to effect peace in the Middle East have been accompanied by a series of books analyzing complex relationships in the region and laying out his proposed plans for peace negotiations and agreements followed by interviews in which many of his answers refer back to assertions made in his books.[9] The book plays an important part in the discussion of complex relationships of power, serving as the source document for ideas and arguments that can be outlined in speeches and interviews but need a text the size of a book for their full elaboration.

The Quiet Crisis

Udall signed a book contract with New York publisher Holt, Rinehart, and Winston on September 25, 1961. Udall's friend the novelist Wallace Stegner, who in 1954 had published the nonfiction work *Beyond the Hundredth Meridian: John Wesley Powell and the Second Opening of the West*, provided an outline for *The Quiet Crisis* while briefly on the Department of the Interior's payroll in the fall of 1961. For two years, Udall worked with a number of individuals inside and outside the department who drafted sections, added information, and read, commented on, and edited versions. Among them were two of Stegner's students, Harold Gilliam and Donald Moser. Gilliam had published a book, *San Francisco Bay*, in 1957 and was involved in the West Coast's environmental movement. He wrote a twenty-five-page (double-spaced) response to one of Udall's drafts in October 1962. The Sierra Club published Moser's book about Olympic National Park, *The Peninsula*, in 1962. Udall acknowledged the help of Stegner, Gilliam and Moser, Sharon Francis, Oliver LaFarge, Ernest Lefever,

Figure 17. Secretary of the Interior Stewart L. Udall posing outdoors with a copy of his book *The Quiet Crisis*. (Records of the Office of the Secretary of Interior, RG-SU, box 1, Still Picture Branch, National Archives and Records Administration, College Park, Maryland)

William H. Whyte, and Joseph L. Fisher, as well as Alvin M. Josephy Jr., who selected the twelve color and thirty black-and-white pictures that illustrated the text.[10]

Udall thus involved a network of people in the writing of his book, just as Carson did. Carson wrote to researchers for information about their work and for clarification and elaboration of key points. She asked knowledgeable scientists and researchers to vet her work, and she hired individuals to help track down information, conduct interviews, and prepare manuscripts. Experienced and respected as a writer, Carson involved others to make sure her work was scientifically accurate and above the reproach she knew the chemical industry would throw at her and because a series of illnesses, including breast cancer, sapped her intellectual and physical energy.[11] In contrast, Udall, a novice at book writing, involved others early in the process to provide guidance regarding the organization and content of the work.

Almost two years to the day after the book contract was signed, the cover of the trade magazine *Publishers Weekly* announced "The Quiet Crisis is coming November 18th," with the text printed over the photograph of mountains silhouetted against a cloudy sky that would appear on the front dust jacket of the book.[12] The 209-page book opened with a foreword in which Udall underscored the meaning of its title: "America today stands poised on a pinnacle of wealth and power, yet we live in a land of vanishing beauty, of increasing ugliness, of shrinking open space, and of an overall environment that is diminished daily by pollution and noise and blight. . . . This, in brief, is the quiet conservation crisis of the 1960's."[13]

President Kennedy then introduced the book proper, taking Udall's foreword a step further: "Our economic standard of living rises, but our environmental standard of living—our access to nature and respect for it—deteriorates. . . . And the long-run effect will be not only to degrade the quality of the national life but to weaken the foundations of national power."[14]

Fourteen chapters followed; throughout, Udall refers to what he calls "The Myth of Superabundance," the idea that the nation's natural resources were limitless, as indeed they seemed to the early explorers and settlers, who encountered abundant forests, wetlands, petroleum deposits, and fish and game. The first twelve chapters present a chronological unfolding of developments in the American environmental movement, told through the ideas and actions of key individuals.

Chapter 1, "The Land Wisdom of the Indians," asserts that Native Americans were emotionally attached to Mother Earth while white settlers wanted to own the land. The second chapter, "The Birth of a Land Policy: Thomas Jefferson," focuses on the relationship of land ownership and the democratic right to vote in the early Republic and the decisions about land acquisition and distribution made during Jefferson's era. In "The White

Indians: Daniel Boone, Jed Smith, and the Mountain Men," Udall writes that
the early trappers and traders presaged the later large-scale excesses of lumber-
men and cattle barons and buffalo-hide dealers and exemplified an ongoing
American "combination of love for the land and the practical urge to exploit it
shortsightedly for profit."[15] Chapter 4, "The Stir of Conscience: Thoreau and
the Naturalists," discusses botanist William Bartram, ornithologist John James
Audubon, historian Francis Parkman, transcendental philosopher Ralph
Waldo Emerson, and writer Henry David Thoreau, who together "began the
development of an American land-consciousness and set in motion a salutary
countercurrent of ideas against the raider spirit of their era."[16] In "The Raid on
Resources," Udall chronicles the demise of virgin forests, the plundering of oil
and natural gas deposits, the attempts to farm arid land, the extinction of the
passenger pigeon and extermination of the buffalo and bison. The next chapter
extols "The Beginning of Wisdom: George Perkins Marsh," whose 1864 book,
Man and Nature, warned that human activity had a profoundly negative impact
on the earth and its resources. Chapter 7 turns to "The Beginning of Action:
Carl Schurz and John Wesley Powell," wherein Udall points out that Schurz,
better known for his efforts to reform the civil service system, worked as secretary
of the interior to save what was left of the forests in the late 1870s and to act on
the department's (John Wesley) Powell Survey of the arid regions west of the
ninety-eighth meridian, which documented the damage done to land and water
resources by ranching and farming.

    Udall devoted chapter 8 to one of his heroes, Gifford Pinchot, who in 1905
effected the transfer of the forest reserves from the Department of Interior to
his own newly renamed bureau in the Department of Agriculture, the Forest
Service. As chief forester, Pinchot balanced preservation and enjoyment of the
forests with commercial uses of woodland resources. Turning to the preservation
of wilderness and the creation of national parks, Udall applauds the efforts
of Sierra Club founder John Muir, who rallied support for, among others,
Yosemite and Mesa Verde National Parks and what would become Grand
Canyon National Park. Chapter 10, "Men Must Act: The Roosevelts and
Politics," recounts Theodore Roosevelt's use of his expanding presidential
power to set aside wilderness areas and protect natural resources and wildlife.
Udall then discusses President Franklin Delano Roosevelt's broad efforts to
reclaim damaged lands and resources through such sweeping government
initiatives as the Tennessee Valley Authority and the Civilian Conservation
Corps. In chapter 11, "Individual Action: Organizers and Philanthropists,"
Udall touches on monetary and land donations by John D. Rockefeller Jr. and
the formation and activities of organizations such as the National Audubon,
Wilderness, and National Geographic Societies; the National Wildlife Federa-
tion; and the Nature Conservancy. Udall tells the story of "Cities in Trouble"
by drawing on the example of urban planner Frederick Law Olmsted, designer

of New York's Central Park and promoter of green areas in cities. His penulti-
mate chapter, "Conservation and the Future," turns attention to science as the
provider of new technologies that help ensure continued abundance, even as
they generate residual problems such as trash and pollution. "Intoxicated with
the power to manipulate nature," Udall writes, "some misguided men have
produced a rationale to replace the Myth of Superabundance. It might be
called the "Myth of Scientific Supremacy," for it rests on the rationalization
that the scientists can fix everything tomorrow."[17] He promotes, among other
things, President Kennedy's proposed Land and Water Conservation Fund,
the expansion of protected wilderness areas and parks, the advantages of cost-
benefit analysis when considering new projects, and a cooperative international
approach to the management of the oceans.

The final chapter, only three pages long, urges Americans to value the
natural world and their place in it: "Henry Thoreau would scoff at the notion
that the Gross National Product should be the chief index to the state of the
nation, or that automobile sales or figures on consumer consumption reveal
anything significant about the authentic art of living. . . . To those who complain
of the complexity of modern life, he might reply, 'If you want inner peace find
it in solitude, not speed, and if you would find yourself, look to the land from
which you came and to which you go.'"[18]

Promotion and Reception

Unlike many authors, the secretary of the interior had no time to "go on the
road" to do a series of readings and signings to promote sales.[19] Nevertheless,
Udall took an active role in making his book known to key individuals, as is
apparent from the correspondence, about seventy letters, in a folder in the
Udall papers labeled "fan mail." The label suggests letters written by individuals
unknown by the addressee expressing their admiration from afar, and about a
quarter of these letters fall into that category. The remaining are more accurately
described as thank-you notes acknowledging receipt of a copy of *The Quiet Crisis*
inscribed by its author. A few are from family members, but the rest are from
influential individuals in publishing, broadcasting, politics, and environmental
organizations. Several of these writers admit to not having read the book in its
entirety before penning a thank-you note.[20]

Udall sent inscribed copies to and received letters from Senators Wayne
Morse and Maurine Neuberger of Oregon; Senator Hubert Humphrey of
Minnesota; Senator Lee Metcalf of Montana, who inserted a review from the
Montana State University student newspaper in the *Congressional Record*; Senator
George McGovern of South Dakota, who inserted three pages into the *Congres-
sional Record* regarding the book; and Congressman Thomas P. Gill of Hawaii.
With the insertions in the *Congressional Record*, Udall's efforts to promote his

thinking reached beyond the elected officials he contacted directly. Udall also sent autographed books to representatives of print and broadcast media, including Brooks Atkinson and Max Frankel of the *New York Times*; Ron Heady of the *Kansas City Star*; Bert Hanna of the *Denver Post*; Jonathan Daniels of the *Raleigh (North Carolina) News and Observer*; Don Moser of *Life* magazine (who had helped in the early stages of the manuscript); Newton Minow of the *Encyclopedia Britannica*; and Howard K. Smith of ABC. Nature writers Sigurd Olson and Joseph Wood Krutch; economist John Kenneth Galbraith; political science professor Clinton Rossiter; and historian Arthur Schlesinger Jr. (on assignment at the White House) also received autographed copies. Some of the gift books were used as review copies by their recipients; for example, Krutch wrote a review for the *New York Herald Tribune* and the *San Francisco Examiner* and Bert Hanna reviewed the book for the *Denver Post*.[21] As with the insertions in the *Congressional Record*, the reviews and notices in the newspapers spread Udall's words far beyond the individuals who received copies of the book from him.

The November 23, 1963, issue of *Saturday Review* featured a cover photo of a river canyon with the tag line "In this issue: A major preview of Stewart L. Udall's *The Quiet Crisis* and a review by Fairfield Osborn." The preview was an excerpt from the final two chapters of the book, along with five illustrations. The piece ran on pages 19–22 and then jumped to pages 52–54. Osborn's review, sandwiched between on page 39, was titled "Long Live the Wild Land." A positive summary of the book's main points, the review concludes, "How right the author is [in the last two chapters] when he stresses the point that the solution to this problem must not be left to the capacities of science and technology but must be motivated by a general acceptance of a 'land ethic for tomorrow.'"[22]

A week later, in its November 30, 1963, issue, the *New Republic* published a review by that magazine's former publisher Michael Straight titled "The Raid on Resources." Straight, himself a former federal government employee whom President Kennedy had considered for a post in what would become the National Endowment for the Arts, wrote, "Stewart Udall is a leading executive and spokesman of the New Frontier.... Udall shouldered by day the administrative burdens of the Department; in his evenings he set out to trace his lineage as Secretary; and so, to define his own role. *The Quiet Crisis* is the result: an eloquent assessment of the tradition in which he stands."[23]

Most of the reviews that followed were positive, including the more than 65 published in newspapers across the United States. Major newspapers of record such as the *New York Times*, the *Los Angeles Times*, the *Chicago Sun-Times*, and the *Christian Science Monitor* ran reviews, and the Monitor published an excerpt as well. But smaller papers—from Schenectady, New York, to Sulphur Springs, Texas—also published reviews of *The Quiet Crisis*. Almost all appeared the same month it was officially published, with a few in December 1963 and January 1964. And almost all were positive. Fairly typical was Hal Borland's piece in the

Figure 18. Secretary of the Interior Stewart L. Udall surrounded by books and papers in his office. (Records of the Office of the Secretary of Interior, RG-SU, box 1, Still Picture Branch, National Archives and Records Administration, College Park, Maryland)

New York Times Book Review: "Because Mr. Udall is Secretary of the Interior, this book must be taken as a statement of public policy. As such, it should command the respect of most conservationists. . . . Beyond policy aspects, however, this is one of the best summaries yet written of the long story of the relationship between Americans and the land."[24]

Within three months of its publication, the book was on *Publishers Weekly*'s best-seller list, along with other nonfiction titles such as President Kennedy's *Profiles in Courage* and Victor Lasky's biography *John Fitzgerald Kennedy: The Man and the Myth*; Jessica Mitford's *The American Way of Death*; and President Dwight D. Eisenhower's *Mandate for Change*. On the fiction list were Mary McCarthy's *The Group*, John LeCarre's *The Spy Who Came in from the Cold*, and Ian Fleming's *On Her Majesty's Secret Service*.[25]

The reviews in scholarly journals appeared over a two-year span and were more measured in their evaluations. The journals of the American Association for the Advancement of Science, the American Institute of Architects, the Ecological Society of America, and the American Ornithologists Union all published reviews, as did the *Journal of Soil & Water Conservation*, the *Journal of Range Management*, *Natural Resources Journal*, *Geographical Review*, and the *Journal of Wildlife Management*. A review by Donald J. Zinn, a zoologist at the University of Rhode Island, was the first in the scholarly journals, appearing in *Science* in January of 1964: "This history is sound, eloquent, vigorous, and excellently written. . . . [It] should be required reading for all conservationists."[26] A few months later, E. William Anderson of the U.S. Soil Conservation Service in Portland, Oregon, wrote in the *Journal of Range Management*: "The author's selection of events and people, his succinct analysis of their impact on succeeding developments and the chronology with which he guides the reader through this maze of national history makes the book easy to read and understand. From beginning to end, it is well-written, interesting, and worthwhile." But Anderson faulted Udall for leaving out "the soil conservation district movement."[27] In January 1965, more than a year after the book's release, Robert A. McCabe of the Department of Wildlife Management at the University of Wisconsin–Madison, wanted a more detailed conclusion: "The final chapter, 'Notes on a Land Ethic for Tomorrow,' is a disappointing summation." Nevertheless, he praised the overall effort as "a superbly written book (the last chapter notwith-standing) with a perceptive eye to the salient facets of history which affected conservation."[28] More measured still was William C. Robison's essay in the April 1965 issue of the *Geographical Review*. Robison wrote that the book "should be widely read both in the colleges and by the public at large." But he questioned "the dubious proposition that the American Indian was endowed with an innate wisdom regarding natural resources." He named two influential individuals who deserved space in the book but whom Udall did not discuss, and he faulted Udall for not citing John Ise, who had written books on federal policies regarding oil, forests, and national parks.[29] Just a month later, A. J. Sharp of the University of Tennessee and the University of Michigan Biological Station wrote in the journal *Ecology*: "Well-written and thought-provoking, the book should be read by every citizen regardless of his profession, his politics, or theories of economics."[30]

The Quiet Crisis and the Next Generation

Udall left Interior when Richard Nixon took office as president in 1969. He remained in Washington, D.C., until 1979, when he moved to Phoenix to practice law.[31] He continued to write, and in 1988 he published *The Quiet Crisis and the Next Generation*, which included the original 1963 text updated with an additional eight chapters, titled:

> Ecology to the Forefront: Rachel Carson and *Silent Spring*
> The Flowering of Environmental Activism: David Brower and the Rise of the Sierra Club
> Howard Zahniser and the Fight for Wilderness
> Science, Law, and Environmental Reform: The Environmental Defense Fund Blazes a Trail
> Widening the Circle of Environmental Awareness: Ralph Nader, Barry Commoner, Paul Ehrlich, and Earth Day
> Confronting the Problem of Human Restraint
> Encounter with Reaganism
> Ecology and the Future.[32]

In his introduction to the new edition, Udall expressed regret for asserting in the first edition his belief in the power of science and technology as positive forces for resource conservation. His understanding of the consequences of science policy had broadened throughout the years as he went to court seeking redress for those affected by the health consequences of uranium mining and nuclear bomb testing. In 1984, he was one of the attorneys defending Navajo miners in a case wherein the U.S. District Court in Arizona, deciding it had no jurisdiction over the matter, recommended the plaintiffs work to have Congress pass a law instead. That federal statute, the Radiation Exposure Compensation Act, would be signed into law in 1990. But as early as 1979 Udall had excoriated the Carter administration for suppressing evidence in an attempt to avoid paying victims, when he testified at a congressional hearing regarding an early version of the act.[33] His realization that the government would deliberately lie changed him, as historian David Emmons would note in his foreword to Udall's *Forgotten Founders: Rethinking the History of the Old West*, issued in 2003. The optimism regarding the government's will and ability to intercede on the environment's behalf so apparent in *The Quiet Crisis* had faded.[34] His 1994 book, *The Myths of*

August: A Personal Exploration of Our Tragic Cold War Affair with the Atom, asserts that Udall's determination to call the government to account began in 1979, when he left Washington, D.C., and returned to Arizona.[35] Udall's somewhat positive assessment of science and technology in his 1963 book, tempered though it was by a warning against the Myth of Scientific Supremacy, had changed. From Udall's perspective, the twenty-five years between the first and second editions of his book were characterized by the unmasking of Big Science and the slow but steady progress of the environmental movement. Even the systematic thwarting of laws and regulations designed to protect the environment and conserve natural resources characteristic of the Reagan era could be understood as a net gain for the environmental movement, if not for the environment, since it had inspired a substantial increase in the membership rolls of existing conservation organizations and the founding of new ones. In 1963, Udall had sounded a note of optimism regarding science and technology as the keys to solving the problem of resource depletion. In 1988, his optimism lay instead with activists who worked to hold government and industry accountable while at the same time heightening the public's awareness of the need to save fragile ecosystems. He was especially heartened by mass spectacles such as Earth Day, which he saw as evidence that the environmental movement was winning the hearts and minds of everyday Americans.[36]

The picture of history through the 1970s and 1980s that Udall drew in *The Next Generation* included a view of his own work as contributing to a positive change in Americans' perceptions of ecology and conservation. In explaining the need for the 1988 edition, he wrote: "Because I had helped orchestrate the beginning of the environmental movement and had participated in many of the disputes that made it a major new political and social cause in this country, friends urged me to update *The Quiet Crisis* in order to share with a new generation the ecological insights that had transformed the conservation cause into a latter-day environmentalism that has become a dominant strain in American thought."[37]

Udall's chapter on Reaganism suggests another motivation for updating *The Quiet Crisis*: the author's disgust at President Ronald Reagan's eight-year-long hostility toward environmental issues. Udall criticized Reagan and his appointees to the Environmental Protection Agency and the Interior Department as committed to supporting business and the so-called free market at the expense of the environment. By Udall's assessment, Reagan's policies led to the waste of oil and other energy resources, deliberate inaction on toxic waste clean-up, and careless disregard for a host of issues from endangered species to nuclear power plant safety. But what Udall objected to most was the way the Reagan administration expressed its disdain for environmental protection and conservation laws and regulations. The administration refused to engage or confront the people's representatives in Congress; instead of introducing new legislation or amendments to existing laws, Udall wrote, it simply refused to fund agencies sufficiently to support their regulatory functions.[38]

But Udall also wanted to acknowledge the impact of Rachel Carson's book in the years since it had awakened people to the dangers of ignorance and complacency. A month after Carson's death in April 1964, Udall published a tribute to her in the *Saturday Review*: "The lyric tone of her prose, the insights she drew from her research, her clear commitment to nature's scheme of things made her a memorable teacher."[39] In 1978 he had tried to get an article he wrote, titled "The Rich Legacy of Rachel Carson," published in *Audubon Magazine*, but the editor rejected the idea on the basis of "overkill."[40] He then tried *Ladies Home Journal*, which also rejected it.[41] Udall understood that Carson had invigorated a latent environmental activism. The first of the additional eight chapters in *The Quiet Crisis and the Next Generation* was devoted to her legacy, citing her work as the catalyst that led to the creation of the Environmental Protection Agency, among other things. Udall argued that her work's greatest impact was its role in what he called "a reorientation of human thought."[42] He invoked Carson's work in six of the remaining chapters.

Although he no longer had the political power of the Interior Department to interest others in his work, he still had the personal power. He acknowledged nine individuals who served as overall critics, including his wife, Lee; his nephew James; and Rachel Carson's editor at Houghton Mifflin, Paul Brooks, whose biography of Carson he also used in his chapter on her. Additionally, he acknowledged those who advised him regarding specific chapters. For example, he thanked Paul Oehser, whom he referred to as "Zahnie's bosom friend," for information used in the seventeenth chapter, on Howard Zahniser's work with the Wilderness Society and for the passage of the Wilderness Act.[43] Both Oehser and Paul Schaefer, vice president of and archivist for the Association for the Protection of the Adirondacks, had suggested that Udall add more material about one of the key founders of the Wilderness Society, Bob Marshall, and both referred to a new book on Marshall by James Glover.[44] Udall revised his section on Marshall and mentioned the book in his acknowledgments for the chapter.[45] Although he no longer had the use of a government agency staff, Udall corresponded and conversed with knowledgeable people as he wrote succeeding drafts of the chapters updating the 1963 work.

Despite the fact that alarm over Reaganism had increased the membership rolls of environmental organizations, there apparently was little commercial interest in an update of a twenty-five-year-old book by an author no longer in a powerful federal government position. A small publisher in Salt Lake City, Gibbs-Smith, published the book, and the print run was considerably smaller than that for the first edition. The book was not widely reviewed; a search of the news files in Lexis-Nexis Academic Universe turned up no reviews in newspapers. In *Audubon Magazine*, in a review titled "Camelot Seems Far Away

Indeed," Frank Graham Jr. evaluated the impact of the first edition and applauded the work of the second, noting that the added chapters are not about historical figures but about Udall's own contemporaries.[46] The *Library Journal* recommended the book in a one-paragraph notice, and reviews appeared in the journals *Landscape Architecture* and *Issues in Science and Technology*, the latter noting Udall's shift from a belief in technological solutions.[47] The WorldCat database shows that just fewer than four hundred libraries own *The Quiet Crisis and the Next Generation*, compared to the more than two thousand libraries that own *The Quiet Crisis*.

Such is the typical fate of subsequent editions. More unusual is the reissue that garners significant interest, such as Richard Dawkins's *The Selfish Gene*.[48] Originally issued in 1976, the thirtieth anniversary edition and a collection of essays, *Richard Dawkins: How a Scientist Changed the Way We Think*, received a lengthy review in the *Times Literary Supplement*, among other periodicals.[49] Reviewer Jerry Coyne referred to *The Selfish Gene* as a "masterpiece" and attributed its status to "the deep thought, the stylish expression, and the sheer self-assurance that lets you know from the outset that you are in the hands of a master."[50] Amazon.com's sales-ranking algorithm puts the second edition of *The Selfish Gene*, published in paperback in 1990, at No. 60,286. The third edition, published in paperback in 2006 with the title *The Selfish Gene: Thirtieth Anniversary Edition—with a New Introduction by the Author*, evidenced much higher sales, with a rank of No. 902.[51] No doubt the coinciding publication of the collection of essays addressing how Dawkins "changed the way we think" helped spur sales of the anniversary edition.

In this case, and perhaps in the case of *Silent Spring*, a provocative thesis by an articulate scientist who writes effectively for a popular audience but also has credibility among his or her scientific peers results in an unbeatable combination that keeps a book alive in the minds of readers. To the extent that Udall's first edition was received as a policy statement wrapped in a history of resource conservation, it was perhaps a bit more fleeting. And to the extent that the 1988 edition was a reissue of the earlier work plus an update that was not even half as long as the original text, the popularity of the second edition was more fleeting still.

A bit more data can help illustrate this point. The Web of Science, a database that tracks bibliographic citations in articles, indicates that as of August 2009, six of the research articles indexed in the database cited *The Quiet Crisis and the Next Generation*, and two of those also cited *The Quiet Crisis*. In contrast, the database yields records for ninety-seven articles (other than book reviews) that cite the 1963 edition. Of those articles, thirty-four were published in 1989 or later and could have cited the 1988 edition but cited the 1963 book instead. A Google Books search was used to try to quantify the books that cite the two editions. The search "Udall 'quiet crisis' 1963" yielded 651 results in August

2009, compared to 88 results for "Udall 'quiet crisis and the next generation' 1988." It is clear that the 1963 edition continues to be known and used by scholars in various disciplines related to environmental history, policy, and science; it is cited today as much for its historical significance as for its content (although of course its historical significance rests at least in part on its content).[52] And in some cases, it is the 1963 portion of the book that is of interest to those citing the 1988 edition.[53] Authors have even cited the 1988 edition as *The Quiet Crisis* rather than by its full title, suggesting that they used the 1988 edition but only for its 1963 text.[54] In another example, an *Outside* magazine article recommended that President Obama read *The Quiet Crisis and the Next Generation*, which it stated was published in 1963.[55] Such mix-ups indicate that many readers do not make the careful distinctions between editions that collectors, bibliographers, and librarians make. From one perspective, it may be enough to know that Udall's work on the history of natural resource conservation and preservation continues to have an impact. From another perspective, however, the different levels of notice that the works have received allow for a consideration of what two books can and cannot do.

Conclusion

In some ways, the success of Udall's first edition may itself have had an impact on the relative lack of notice of the twenty-fifth anniversary update, and not just because the update included all of the text of the original edition. One year after *Silent Spring*, readers were primed for more books that explained their place in the natural world. Udall understood that the timing was right for his own work, not just because he was a federal official with an agenda but also because *Silent Spring* had created a market. A Salt Lake City reporter who interviewed Udall in 2002 wrote, "Udall believes he was able to capitalize on Carson's excellent 'kick-off' by writing his own book and promoting the environment through presidents who were friendly to his policies."[56] The author's prominence as a government official helped generate interest in the first edition. He was no longer such a public person when the revised version entered a marketplace with many more competing books. In the ensuing twenty-five years, numerous books on the environment had appeared, as Udall discussed in his *Next Generation*. A timeline at the back of *Next Generation* lists the 1962 publication of *Silent Spring* as the first important event of what Udall called "The Age of Ecology." Other "significant publications" shown in the timeline are Leo Marx's *The Machine in the Garden: Technology and Pastoral Ideal in America* (1964); Kenneth Boulding's *The Economics of Spaceship Earth* (1966) and *Ecodynamics* (1981); Roderick Nash's *Wilderness and the American Mind* (1967); Garrett Hardin's *The Tragedy of the Commons* (1967); Paul Ehrlich's *The Population Bomb* (1968); *The Cold and the Dark* (1984) by Paul R. Ehrlich, Carl Sagan, Donald Kennedy, and

Walter Orr Roberts, which included variously authored scientific papers on nuclear winter; and twelve more. The acknowledgments at the end of the book list a few additional titles he relied on for specific chapters.[57] The 1970s and 1980s also saw a blossoming of books by professional environmental historians, chronicled by William Cronon.[58] What may be understood as the novelty of *Silent Spring*—not to mention the novelty of an interior secretary capable of writing a lucid and stirring popular account of environmental history—helped propel the 1963 edition of *The Quiet Crisis* to best-seller status. That it also helped launch a long-term book publishing trend that defined the print culture of the environmental movement and its history during the latter half of the twentieth century may at least partly account for the quiet reception of *The Quiet Crisis and the Next Generation*. In a deepening sea of books about pollution, wildlife, conservation, preservation, the environment, ecology, endangered species, wilderness, natural resources, energy, and related aspects such as women's involvement in the green movement, the *Next Generation* could not make as much of a splash as its predecessor. By the time *The Next Generation* came out, the popular and scholarly print culture of the environmental movement had become established and the broad topic of ecology was more developed. *The Next Generation* was not so much reviewed immediately as it was cited over time.

Simply put, a first edition can make an impact—can make history—in a way subsequent editions cannot, by virtue of being the author's first extended statement on the topic. Even when a subsequent edition, such as the third edition marking the thirtieth anniversary of *The Selfish Gene*, makes a splash, it depends on the publication of the first edition. The subsequent edition launches activities including not only publication but also the attendant marketing, promoting, and reviewing processes that attract attention, creating its own life but also, at least in some cases, injecting new life into its old version. The first edition speaks to one generation, the subsequent edition to another, as Udall's 1988 title indicates. In the print culture of the environmental movement, an influential first edition spawns political action and inspires further publications. A subsequent edition has some chance of doing the same, but generally on a smaller scale, as it works in the shadow of its more groundbreaking predecessor. A book may have a life of its own, beyond the author's vision for it, but a book and its later expanded version can lead separate as well as intertwining lives, one as a first edition and the other as a subsequent work that operates in a milieu at least partly created by its own earlier iteration. It may seem obvious, but it is worth stating outright nonetheless: a subsequent edition of a book enters a context different from the one its first edition entered, and one of the differences, of course, is the prior existence of the first edition.[59] The context is also changed by other books that exist in explicit or implicit response to that first edition, a feature that those books have in common with the subsequent edition. When *The Quiet Crisis* appeared in 1963, nothing else quite like it existed. When *The*

Quiet Crisis and the Next Generation appeared in 1988, something else very much like it did.

Notes

The author thanks Fernando Elichirigoity for his insightful suggestions on an early version of this essay; audience members who asked questions and made suggestions when I presented early work on Udall at the Culture of Print in Science, Technology, Engineering, and Medicine conference in Madison, Wisconsin; the editors and anonymous reviewers for their helpful guidance; and University of Arizona archivists Roger Myers and Chrystal Carpenter for assistance with the Stewart L. Udall Papers.

1. Priscilla Coit Murphy, *What a Book Can Do: The Publication and Reception of "Silent Spring"* (Amherst: University of Massachusetts Press, 2005), 2.

2. "Stewart Udall," *Contemporary Authors Online* (Detroit: Gale, 2010); *Biography Resource Center* (Farmington Hills, Mich.: Gale, 2010); Lewis L. Gould, *Lady Bird Johnson and the Environment* (Lawrence: University Press of Kansas, 1988), 41–42; Anne Becher, "Udall, Stewart," *American Environmental Leaders* (Santa Barbara, Calif.: ABC-Clio, 2000), 821–24; Peter Wild, "Working from Within: Stewart Udall Finds Conservation Good Politics," *Pioneer Conservationists of Western America* (Missoula, Mont.: Mountain Press, 1979), 173–83; L. Boyd Finch, *Legacies of Camelot: Stewart and Lee Udall, American Culture and the Arts* (Norman: University of Oklahoma Press, 2008); Morris K. Udall and Stewart L. Udall Foundation, "About Stewart L. Udall," http://www.udall.gov/AboutSLUdall/About SLUdall.aspx; Keith Schneider and Cornelia Dean, "Stewart L. Udall, Conservationist in Kennedy and Johnson Cabinets, Dies at 90," *New York Times*, March 20, 2010, http://www.nytimes.com/2010/03/21/nyregion/21udall.html.

3. Robert M. Utley and Barry Mackintosh, *The Department of Everything Else: Highlights of Interior History* (National Park Service, 1989), http://www.nps.gov/history/history/online_books/utley-mackintosh/interior13.htm.

4. Thomas G. Smith, "John Kennedy, Stewart Udall, and New Frontier Conservation," *Pacific Historical Review* 64, no. 3 (August 1995): 336.

5. Linda Lear, *Rachel Carson: Witness for Nature* (New York: Henry Holt, 1997), 404–6; Mark Hamilton Lytle, *The Gentle Subversive: Rachel Carson, "Silent Spring," and the Rise of the Environmental Movement* (Oxford: Oxford University Press, 2007), 164.

6. Jeffrey K. Tulis, *The Rhetorical Presidency* (Princeton, N.J.: Princeton University Press, 1987).

7. For example, see Karlyn Kohrs Campbell and Kathleen Hall Jamieson, *Deeds Done in Words: Presidential Rhetoric and the Genres of Governance* (Chicago: University of Chicago Press, 1990).

8. Michiko Kakutani, "The Politics of Prose," *New York Times*, April 22, 2007, http://www.nytimes.com/2007/04/22/books/22kaku.html.

9. One example is Terry Gross's interview on the *Fresh Air* radio program, "Jimmy Carter Offers a Peace Plan 'That Will Work,'" January 27, 2009, http://www.npr.org/templates/story/story.php?storyId=99875313.

10. [Contract], September 25, 1961, box 213, folder 1, and Harold Gilliam, "*The Quiet Crisis*: Analysis," October 1962, box 213, folder 2, Papers of Stewart L. Udall,

1950–77, courtesy of University of Arizona Library, Special Collections (hereafter Udall Papers); Stewart L. Udall, *The Quiet Crisis* (New York: Holt, Rinehart, and Winston, 1963), 193; Finch, *Legacies of Camelot*, 65–76.

11. Lear, *Rachel Carson*, 400–402; Murphy, *What a Book Can Do*, 45–46; Lytle, *Gentle Subversive*, 161–64.

12. *Publishers Weekly*, September 23, 1963, Udall Papers, box 218, "Album of Review Clippings."

13. Udall, *Quiet Crisis*, viii.

14. John F. Kennedy, introduction to Udall, *Quiet Crisis*, xiii.

15. Udall, *Quiet Crisis*, 37.

16. Ibid., 53.

17. Ibid., 178.

18. Ibid., 190.

19. Author's telephone interview with Stewart L. Udall, November 24, 2009. I am grateful to the Honorable Tom Udall for his help arranging this interview with his father.

20. An example, on National Audubon Society letterhead, is from Carl W. Buchheister to Stewart L. Udall, November 5, 1963, Udall Papers, box 213, folder 3.

21. Wayne Morse to Udall, October 28, 1963; Maurine B. Neuberger to Udall, October 29, 1963; Hubert H. Humphrey to Udall, December 4, 1963; Lee Metcalf to Udall, December 17, 1963; George McGovern to Udall, November 20, 1963; Thomas P. Gill to Udall, November 6, 1963; Brooks Atkinson to Udall, November 2 and 4, 1963; Max Frankel to Udall, [n.d., stamped at Interior December 6, 1963]; Ron Heady to Udall, November 3, 1963; Bert Hanna to Udall, November 2, 1963; Jonathan Daniels to Udall, November 1, 1963; Don Moser to Udall, November 18, 1963; Newton N. Minow to Udall, November 21, 1963; Howard K. Smith to Udall, November 5, 1963; Sigurd F. Olson to Udall, November 23, 1963; Joseph Wood Krutch to Udall, October 9, 1963; John Kenneth Galbraith to Udall, November 5, 1963; Clinton Rossiter to Udall, November 4, 1963; Arthur Schlesinger, Jr. to Udall, October 28, 1963; Udall Papers, box 213, folder 3.

22. Stewart L. Udall, "The Quiet Crisis," *Saturday Review*, November 23, 1963, 19–22, 53–54; Fairfield Osborne, "Long Live the Wild Land," *Saturday Review*, November 23, 1963, 39, clippings in Udall Papers, box 218.

23. Michael Straight, "The Raid on Resources," *New Republic*, November 30, 1963, [n.p.], clipping in Udall Papers, box 218.

24. Hal Borland, *New York Times Book Review*, December 8, 1963, [n.p.], clipping in Udall Papers, box 218.

25. "Best Sellers of the Week," *Publishers Weekly*, February 3, 1964, 160, clipping in Udall Papers, box 218.

26. Donald J. Zinn, "Review: Conservation," *Science* 143, no. 3605 (January 31, 1964): 458.

27. E. William Anderson, "Review," *Journal of Range Management* 17, no. 3 (May 1964): 158.

28. Robert A. McCabe, "Review," *Journal of Wildlife Management* 29, no. 1 (January 1965): 226–27.

29. William C. Robison, "Review," *Geographical Review* 55, no. 2 (April 1965): 308–10.

30. A. J. Sharp, "Review: A History of Conservation and Philosophy of Resource Use in the U.S.," *Ecology* 46, no. 3 (May 1965): 385–86.

31. University of Arizona Library, "Stewart L. Udall: Career Chronology," http://www.library.arizona.edu/exhibits/sludall/career.htm.

32. Stewart L. Udall, *The Quiet Crisis and the Next Generation* (Salt Lake City: Gibbs-Smith, 1988).

33. National Research Council, *Assessment of the Scientific Information for the Radiation Exposure Screening and Education Program* (Washington, D.C.: National Academies Press, 2005), 17–19; *John N. Begay, et al., Plaintiffs, v. United States of America, Defendant; Phillip Anderson, et al., Plaintiffs, v. United States of America, Defendant*, Nos. Civ. 80-982 Pct. WPC, Civ. 81-1057 Pct. WPC (Consolidated). U.S. District Court for the District of Arizona. 591 F. Supp. 991; 1984 U.S. Dist. LEXIS 15044, July 10, 1984; U.S. Senate, Sub-Committee on Health and Scientific Research of the Committee on Labor and Human Resources and the Committee on the Judiciary, *Radiation Exposure Compensation Act of 1979* (HRG-1980-LHR-0049. Date: June 10, 1980), 71–74. Text in LexisNexis *Congressional Hearings Digital Collection*.

34. David M. Emmons, foreword to Stewart L. Udall, *Forgotten Founders: Rethinking the History of the Old West* (Washington, D.C.: Island Press, 2003), xiii–xv.

35. Stewart L. Udall, *The Myths of August: A Personal Exploration of Our Tragic Cold War Affair with the Atom* (1994; repr., New Brunswick, N.J.: Rutgers University Press, 1998), 3.

36. Udall, *Next Generation*, xvi–xvii, 261–62, 243–44.

37. Ibid., xvii.

38. Ibid., 257–62.

39. Stewart L. Udall, "The Legacy of Rachel Carson," *Saturday Review*, May 16, 1964, 23.

40. Les Line to Udall, September 29, 1978, Udall Papers, box 234.

41. Lenore Hershey to Udall, October 13, 1978, Udall Papers, box 234.

42. Udall, *Next Generation*, 203.

43. Ibid., 279–81.

44. Paul H. Oehser to Stewart L. Udall, November 24, 1986; Paul Schaefer to Udall, February 14, 1987; both typed letters in the Udall Papers, box 234.

45. Udall, *Next Generation*, 280.

46. Frank Graham Jr., "Camelot Seems Far Away Indeed," *Audubon Magazine* 91 (January 1989): 115–17.

47. Michael Rogers, "The Quiet Crisis and the Next Generation," *Library Journal* 116, no. 16 (October 1, 1991): 85; Ed Marston, "The Education of Stewart Udall," *Issues in Science and Technology* 5, no. 3 (Spring 1989): 106–9; and E. P. Oberholtzer, "The Quiet Crisis and the Next Generation—Udall, S. L.," *Landscape Architecture* 80, no. 1 (January 1990): 101.

48. Richard Dawkins, *The Selfish Gene* (New York: Oxford University Press, 1976). I am grateful to James A. Secord for this observation and for providing this example.

49. Richard Dawkins, *The Selfish Gene: Thirtieth Anniversary Edition—with a New Introduction by the Author* (New York: Oxford University Press, 2006); and Alan Grafen and Mark Ridley, *Richard Dawkins: How a Scientist Changed the Way We Think: Reflections by Scientists, Writers, and Philosophers* (New York: Oxford University Press, 2006).

50. Jerry A. Coyne, "Thirty Years of the Selfish Gene," *Times Literary Supplement*, June 16, 2006, 7–9.

51. Sales rankings are shown on the amazon.com Web page describing the book. The 1990 edition is at http://www.amazon.com/Selfish-Gene-Richard-Dawkins/ dp/0192860925; the 2006 edition is at http://www.amazon.com/Selfish-Gene-Anniversary-Introduction/dp/0199291152 (accessed August 14, 2009). Amazon sales rankings change dynamically as books sell, so these exact figures may never appear again, but they do give a sense of the relative positions of the two books.

52. Byron E. Pearson points out that the book "occupies a prominent position among the events of the 1960s that, historians argue, helped facilitate the emergence of modern environmentalism" in his *Still the Wild River Runs: Congress, the Sierra Club, and the Fight to Save Grand Canyon* (Tucson: University of Arizona Press, 2002), xvi.

53. One example involves a reference to President Kennedy's preface to the 1963 edition, which was included in the 1988 edition, cited by Henry R. Richmond, "From Sea to Shining Sea: Manifest Destiny and the National Land Use Dilemma," *Pace Law Review* 13, no. 2 (1993): 327n2.

54. Citations giving the 1963 title but the 1988 publication date include Robert Paehlke, "Udall, Stewart L.," in *Conservation and Environmentalism: An Encyclopedia*, ed. Paehlke (New York: Garland, 1995), 649; Charles Davis, ed., *Western Public Lands and Environmental Politics* (Boulder: Westview, 1997), 8; and Robert E. Krebs, ed., *The Basics of Earth Science* (Westport, Conn.: Greenwood, 2003), 169.

55. Elizabeth Hightower, "Code Green: Obama's Bedside Cabinet," *Outside* 34, no. 3 (March 1, 2009): 34.

56. Dennis Lythgoe, "Stewart Udall: Troubled Optimist," *Deseret News*, November 17, 2002, E01, accessed via Lexis-Nexis Academic Universe.

57. Udall, *Next Generation*, 272–74, 279–81.

58. William Cronon, "The Uses of Environmental History," *Environmental History Review* 17, no. 3 (Fall 1993): 1–22.

59. This is not meant to deny that changing social, religious, and political contexts also shape the reception of texts over time, as Aileen Fyfe has documented in "The Reception of William Paley's *Natural Theology* in the University of Cambridge," *British Journal for the History of Science* 30, no. 3 (1997): 321–35.

Note on Sources

FLORENCE C. HSIA

The history of print culture is a house of many rooms, one built since the 1950s on such disparate foundations as "histoire du livre" in the Annales tradition,[1] studies of the book trade and literary marketplace as a form of "history from below,"[2] Anglo-American analytical bibliography and textual criticism as reoriented toward the historical conditions and social processes of textual transmission,[3] theoretical reconceptualizations of authorship and readership,[4] and meditations on the significance of media technology for human perception and communication.[5] But it was *The Printing Press as an Agent of Change* (1979) that catalyzed scholarly attention to print culture's import for the history of science, thanks to the hat trick of cultural watersheds — Renaissance, Reformation, and Scientific Revolution — Elizabeth Eisenstein sought to explain in terms of the advent of print in early modern Europe.[6]

Scholars have found much to debate in Eisenstein's claims for the relative stability and availability of typographic materials in comparison with scribal artifacts, as well as in the larger conclusions to be drawn from such characterizations. Bruno Latour sees in the astronomical books and observations recorded on preprinted forms that Tycho Brahe supposedly accumulated at Uraniborg a feedback mechanism ("immutable mobiles") key to the authority of modern science, but Adrian Johns peers into the murky depths of the early modern English book trade to find Boyle, Hooke, Halley, Flamsteed, and Newton wrestling with issues of piracy and plagiarism, while James Secord surveys the complex landscape of early Victorian publishing to reveal the subtle topography of reader response to Robert Chambers's anonymously published *Vestiges of the Natural History of Creation* (1844).[7] A growing number of thematic volumes, essays in reference works, and major synthetic studies suggest the extent to which issues of textual production and consumption in the age of print have become standard fare for the field.[8]

If there is any trend in this burgeoning literature, it is toward complicating the categories of author and reader, script and print. Students of scientific authorship, for instance, have found in economies of credit and attribution under various conditions of print production a wide range of textual producers (writers, copyists, editors, censors, publishers, printers, booksellers, reviewers, anthologizers, literary executors) with sometimes overlapping, anonymous, collective, or institutional responsibilities.[9]

Attention to the print object as physical artifact at the level of paper type, format, and bindings (*mise-en-livre*) as well as layout, type, ornament, and illustration (*mise-en-page*), together with title pages, tables, indices, booksellers' lists, and other paratextual material has highlighted the mediation of many participants in the production process, however difficult and sometimes impossible they may be to identify. The interplay between image and text in the context of print culture has generated an especially large body of literature.[10] Genre has likewise proven a fruitful perspective, whether in dealing with generic forms that bridge scribal and typographic cultures (textbooks, commentaries, commonplace books, encyclopedias) or those born with or substantially transformed by the age of print. Of the latter, journals and other serial publications have attracted considerable interest.[11]

Finally, readership studies display an impressively wide range of approaches. Deciphering marginalia and reconstructing note-taking practices have offered historically sensitive insights into reading experience both individual and collective, while tracing provenance, editions, and print runs to the diverse social spaces in which reading takes place—be they classrooms or coffeehouses, drawing rooms or diaries, learned societies, libraries, or linguistic communities—has led to the mapping of "geographies of reading" (Secord, 2000), "geographies of reception" (Rupke, 1999), and patterns of translation across natural, vehicular, and technical language groups.[12]

Notes

1. Febvre and Martin (1958) 1976; Bollème et al. 1965, 1970; Martin et al. 1982–86; Chartier 1987.

2. Davis 1975; Darnton 1979, 1982a, 1982b.

3. McKenzie 1969, 1981, 1986; McGann 1983.

4. Barthes (1967) 1977; Foucault (1969) 1977; Jauss (1969) 1982; Iser (1972) 1974; Genette (1987) 1997; Chartier (1992) 1994.

5. McLuhan 1962; Ong 1958, 1967, 1982.

6. Eisenstein 1979, 1983.

7. Latour 1987, 1990; Law 1986; Johns 1998; Secord 2000.

8. McNally 1987; Besson 1990; Connor et al. 1995; Myers and Harris 1998; Hessen-bruch 2000; Hunter 2000; Frasca-Spada and Jardine 2000; Topham 2000; Topham et al. 2000; Grafton, Eisenstein, and Johns 2002; Porter 2003; Rosenberg 2003; Rosenberg et al. 2003; Blair, Topham, and Daston 2004; Henson et al. 2004; Park and Daston 2006.

9. Pardo Tomás 1991; Lowood and Rider 1994; Iliffe 1995; Hunter 1998; Johns 1998; Topham 1998; Secord 2000; Kusukawa 2000; Fyfe 2002; Biagioli and Galison 2003; Hellyer 2005; Hunter 2007; Lightman 2007; Fyfe and Lightman 2007.

10. Rudwick 1976, 1992; Knight 1977; Dennis 1989; Harwood 1989; Duncan 1991; Edgerton 1991; Winkler and Van Helden 1992, 1993; Mazzolini 1993; Blum 1993; Baigrie 1996; Jones and Galison 1998; Rider 1998; Shea 2000; Martin et al. 2000; Pang 2002; Freedberg 2002; Lefèvre, Renn, and Schoepflin 2003; Kaiser 2005; Remmert (2005) 2011; Kusukawa and Maclean 2006; Baldasso 2006; Smith 2006; Mosley 2007; Daston and Galison 2007; Blair 2010; Eddy 2010; Kusukawa 2012.

11. Ellegård 1958; Kronick 1976, 1991, 2004; Meadows 1980; Sheets-Pyenson 1981, 1985; Hufbauer 1982; Gascoigne 1985; Bazerman 1988; Bynum, Lock, and Porter 1992; Howsam 1992, 2000; Blair 1992, 1997, 2010; Ausejo and Hormigón 1993; Crosland 1994; Kusukawa 1995; Allen 1996; Frasca-Spada and Jardine 2000; Johns 2000; Topham 2000; Guédon 2001; Schulten 2001; Yeo 2001; Gross, Harmon, and Reidy 2002; Cantor and Shuttleworth 2004; Cantor et al. 2004; Ogilvie 2006; Campi et al. 2008; Hsia 2009; Kohlstedt 2010; Margócsy 2010.

12. Chartier and Corsi 1996; Engelfriet 1998; Topham 1998; Rupke 1999, 2000; Secord 2000; Mandelbrote 2000; Overmier 2000; Montgomery 2000; Wright 2000; Lackner, Amelung, and Kurtz 2001; Vande Walle and Kasaya 2001; Gingerich 2002; Blair 2003, 2010; Golvers 2003; Johns 2003; Blair, Topham, and Daston 2004; Lackner and Vittinghoff 2004; Livingstone 2005; Johns 2006; Burke and Hsia 2007; Duris 2008.

Works Cited

Allen, David E. 1996. "The Struggle for Specialist Journals: Natural History in the British Periodicals Market in the First Half of the Nineteenth Century." *Archives of Natural History* 23 (1): 107–23.

Ausejo, Elena, and Mariano Hormigón, eds. 1993. *Messengers of Mathematics: European Mathematical Journals (1800–1946)*. Madrid: Siglo XX.

Baigrie, Brian S., ed. 1996. *Picturing Knowledge: Historical and Philosophical Problems Concerning the Use of Art in Science*. Toronto: University of Toronto Press.

Baldasso, Renzo. 2006. "The Role of Visual Representation in the Scientific Revolution: A Historiographic Inquiry." *Centaurus* 48 (2): 69–88.

Barthes, Roland. (1967) 1977. "The Death of the Author." *Aspen* 5–6. Trans. by Richard Howard. In *Image, Music, Text*. Ed. and trans. by Stephen Heath, 142–48. New York: Hill and Wang.

Bazerman, Charles. 1988. *Shaping Written Knowledge: The Genre and Activity of the Experimental Article in Science*. Madison: University of Wisconsin Press.

226FLORENCE C. HSIA

Besson, Alain, ed. 1990. *Thornton's Medical Books, Libraries, and Collectors: A Study of Bibliography and the Book Trade in Relation to the Medical Sciences.* 3rd rev. ed. Aldershot, Hants, England: Gower.

Biagioli, Mario, and Peter Galison, eds. 2003. *Scientific Authorship: Credit and Intellectual Property in Science.* New York: Routledge.

Blair, Ann. 1992. "Humanist Methods in Natural Philosophy: The Commonplace Book." *Journal of the History of Ideas* 53 (4): 541–51.

———. 1997. *The Theater of Nature: Jean Bodin and Renaissance Science.* Princeton, N.J.: Princeton University Press.

———. 2003. "Reading Strategies for Coping with Information Overload ca. 1550–1700." *Journal of the History of Ideas* 64 (1): 11–28.

———. 2010. *Too Much to Know: Managing Scholarly Information before the Modern Age.* New Haven: Yale University Press.

Blair, Ann, Jonathan R. Topham, and Lorraine Daston. 2004. "Focus: Scientific Readers." *Isis* 95 (3): 420–48.

Blum, Ann Shelby. 1993. *Picturing Nature: American Nineteenth-Century Zoological Illustration.* Princeton, N.J.: Princeton University Press.

Bollème, Geneviève, et al., eds. 1965, 1970. *Livre et société dans la France du XVIIIe siècle.* 2 vols. Paris: Mouton et cie.

Burke, Peter, and R. Po-chia Hsia, eds. 2007. *Cultural Translation in Early Modern Europe.* New York: Cambridge University Press.

Bynum, W. F., Stephen Lock, and Roy Porter, eds. 1992. *Medical Journals and Medical Knowledge: Historical Essays.* London: Routledge.

Campi, Emidio, et al., eds. 2008. *Scholarly Knowledge: Textbooks in Early Modern Europe.* Geneva: Librairie Droz.

Cantor, Geoffrey, and Sally Shuttleworth, eds. 2004. *Science Serialized: Representation of the Sciences in Nineteenth-Century Periodicals.* Cambridge, Mass.: MIT Press.

Cantor, Geoffrey, et al. 2004. *Science in the Nineteenth-Century Periodical: Reading the Magazine of Nature.* Cambridge: Cambridge University Press.

Chartier, Roger. 1987. *The Cultural Uses of Print in Early Modern France.* Trans. by Lydia G. Cochrane. Princeton, N.J.: Princeton University Press.

———. 1992 (1994). *L'ordre des livres: Lecteurs, auteurs, bibliothèques en Europe entre XIVe et XVIIIe siècle.* Aix-en-Provence: Alinea. *The Order of Books: Readers, Authors, and Libraries in Europe between the Fourteenth and Eighteenth Centuries.* Trans. by Lydia G. Cochrane. Stanford, Calif.: Stanford University Press.

Chartier, Roger, and Pietro Corsi, eds. 1996. *Sciences et langues en Europe.* Paris: École des hautes études en sciences sociales.

Connor, Jennifer J., et al. 1995. "Book Culture and Medicine." *Canadian Bulletin of Medical History / Bulletin canadien d'histoire de la médicine* [Special issue] 12 (2): 203–445.

Crosland, Maurice P. 1994. *In the Shadow of Lavoisier: The "Annales de chimie" and the Establishment of a New Science.* [Oxford]: British Society for the History of Science.

Darnton, Robert. 1979. *The Business of Enlightenment: A Publishing History of the "Encyclopédie," 1775–1800.* Cambridge: Belknap Press of Harvard University Press.

———. 1982a. *The Literary Underground of the Old Regime.* Cambridge, Mass.: Harvard University Press.

———. 1982b. "What Is the History of Books?" *Daedalus* 111 (3): 65–83.

Daston, Lorraine, and Peter Galison. 2007. *Objectivity*. New York: Zone.

Davis, Natalie Zemon. 1975. *Society and Culture in Early Modern France*. Stanford, Calif.: Stanford University Press.

Dennis, Michael Aaron. 1989. "Graphic Understanding: Instruments and Interpretation in Robert Hooke's *Micrographia*." *Science in Context* 3 (2): 309–64.

Duncan, Alistair. 1991. "The Requirements of Scientific Publishing: The Example of Chemical Illustrations in the Scientific Revolution." *Publishing Research Quarterly* 7 (1): 33–53.

Duris, Pascal, ed. 2008. *Traduire la science: Hier et aujourd'hui*. Pessac: Maison des sciences de l'homme d'Aquitaine.

Eddy, M. D. 2010. "Tools for Reordering: Commonplacing and the Space of Words in Linnaeus's *Philosophia Botanica*." *Intellectual History Review* 21 (2): 227–52.

Edgerton, Samuel Y. 1991. *The Heritage of Giotto's Geometry: Art and Science on the Eve of the Scientific Revolution*. Ithaca, N.Y.: Cornell University Press.

Eisenstein, Elizabeth L. 1979. *The Printing Press as an Agent of Change: Communications and Cultural Transformations in Early Modern Europe*. Cambridge: Cambridge University Press.

———. 1983. *The Printing Revolution in Early Modern Europe*. Cambridge: Cambridge University Press.

Ellegård, Alvar. 1958. *Darwin and the General Reader: The Reception of Darwin's Theory of Evolution in the British Periodical Press, 1859–1872*. Göteborg: Elanders.

Engelfriet, Peter M. 1998. *Euclid in China: The Genesis of the First Chinese Translation of Euclid's "Elements," Books I-VI ("Jihe yuanben," Beijing, 1607) and Its Reception up to 1723*. Leiden: Brill.

Febvre, Lucien, and Henri-Jean Martin. (1958) 1976. *L'apparition du livre*. Paris: Éditions Albin Michel. *The Coming of the Book: The Impact of Printing 1450–1800*. Trans. by David Gerard. London: N.L.B.

Foucault, Michel. (1969) 1977. "Qu'est-ce qu'un auteur?" *Bulletin de la Société française de philosophie* 63 (3): 73–104. "What Is an Author?" in *Language, Counter-Memory, Practice: Selected Essays and Interviews*. Ed. by Donald F. Bouchard, trans. by Donald F. Bouchard and Sherry Simon, 113–38. Ithaca, N.Y.: Cornell University Press.

Frasca-Spada, Marina, and Nicholas Jardine, eds. 2000. *Books and the Sciences in History*. Cambridge: Cambridge University Press.

Freedberg, David. 2002. *The Eye of the Lynx: Galileo, His Friends, and the Beginnings of Modern Natural History*. Chicago: University of Chicago Press.

Fyfe, Aileen. 2002. "Publishing and the Classics: Paley's *Natural Theology* and the Nineteenth-Century Scientific Canon." *Studies in History and Philosophy of Science Part A* 33 (4): 729–51.

Fyfe, Aileen, and Bernard V. Lightman, eds. 2007. *Science in the Marketplace: Nineteenth-Century Sites and Experiences*. Chicago: University of Chicago Press.

Gascoigne, Robert Mortimer. 1985. *A Historical Catalogue of Scientific Periodicals, 1665–1900: With a Survey of their Development*. New York: Garland.

Genette, Gérard. (1987) 1997. *Seuils*. Paris: Éditions du Seuil. *Paratexts: Thresholds of Interpretation*. Trans. by Jane E. Lewin. Cambridge: Cambridge University Press.

Gingerich, Owen. 2002. *An Annotated Census of Copernicus' "De revolutionibus" (Nuremberg, 1543 and Basel, 1566)*. Leiden: Brill.

Golvers, Noël. 2003. *Ferdinand Verbiest, S.J. (1623–1688) and the Chinese Heaven: The Composition of the Astronomical Corpus, Its Diffusion and Reception in the European Republic of Letters.* Leuven: Leuven University Press.

Grafton, Anthony, Elizabeth L. Eisenstein, and Adrian Johns. 2002. "AHR Forum: How Revolutionary Was the Print Revolution?" *American Historical Review* 107 (1): 84–128.

Gross, Alan G., Joseph Harmon, and Michael Reidy. 2002. *Communicating Science: The Scientific Article from the Seventeenth Century to the Present.* Oxford: Oxford University Press.

Guédon, Jean-Claude. 2001. *In Oldenburg's Long Shadow: Librarians, Research Scientists, Publishers, and the Control of Scientific Publishing.* Washington, D.C.: Association of Research Libraries, 2001.

Harwood, John T. 1989. "Rhetoric and Graphics in *Micrographia*." In *Robert Hooke: New Studies.* Ed. by Michael Hunter and Simon Schaffer, 119–47. Woodbridge, England: Boydell.

Hellyer, Marcus. 2005. *Catholic Physics: Jesuit Natural Philosophy in Early Modern Germany.* Notre Dame, Ind.: University of Notre Dame Press.

Henson, Louise, et al., eds. 2004. *Culture and Science in the Nineteenth-Century Media.* Aldershot, Hants, England: Ashgate.

Hessenbruch, Arne, ed. 2000. *Reader's Guide to the History of Science.* London: Fitzroy Dearborn.

Howsam, Leslie. 1992. "Sustained Literary Ventures: The Series in Victorian Book Publishing." *Publishing History* 31: 5–26.

———. 2000. "An Experiment with Science for the Nineteenth-Century Book Trade: The International Scientific Series." *British Journal for the History of Science* 33 (2): 187–207.

Hsia, Florence C. 2009. *Sojourners in a Strange Land: Jesuits and Their Scientific Missions in Late Imperial China.* Chicago: University of Chicago Press.

Hufbauer, Karl. 1982. *The Formation of the German Chemical Community, 1720–1795.* Berkeley: University of California Press.

Hunter, Andrew, ed. 2000. *Thornton and Tully's Scientific Books, Libraries, and Collectors: A Study of Bibliography and the Book Trade in Relation to the History of Science.* 4th ed. Aldershot, Hants, England: Ashgate.

Hunter, Michael, ed. 1998. *Archives of the Scientific Revolution: The Formation and Exchange of Ideas in Seventeenth-Century Europe.* Woodbridge, England: Boydell.

———. 2007. "Robert Boyle and the Early Royal Society: A Reciprocal Exchange in the Making of Baconian Science." *British Journal for the History of Science* 40 (1): 1–23.

Iliffe, Rob. 1995. "'Author-Mongering': The 'Editor' between the Author and Reader in the Late Seventeenth and Early Eighteenth Century." In *The Consumption of Culture, 1600–1800: Image, Object, Text.* Ed. by Ann Bermingham and John Brewer, 166–92. New York: Routledge.

Iser, Wolfgang. (1972) 1974. *Der implizite Leser: Kommunikationsformen des Romans von Bunyan bis Beckett.* Munich: W. Fink. *The Implied Reader: Patterns of Communication in Prose Fiction from Bunyan to Beckett.* Baltimore: Johns Hopkins University Press.

Jauss, Hans Robert. (1969) 1982. *Literaturgeschichte als Provokation der Literaturwissenschaft.* 2nd ed. Konstanz: Universitätsverlag. "Literary History as a Challenge to Literary

Theory." In *Toward an Aesthetic of Reception*. Trans. by Timothy Bahti, 3–45. Minneapolis: University of Minnesota Press.

Johns, Adrian. 1998. *The Nature of the Book: Print and Knowledge in the Making.* Chicago: University of Chicago Press.

———. 2000. "Miscellaneous Methods: Authors, Societies, and Journals in Early Modern England." *British Journal for the History of Science* 33 (2): 159–86.

———. 2003. "Reading and Experiment in the Early Royal Society." In *Reading, Society, and Politics in Early Modern England.* Ed. by Kevin Sharpe and Steven N. Zwicker, 244–71. Cambridge: Cambridge University Press.

———. 2006. "Coffeehouses and Print Shops." In *The Cambridge History of Science.* Vol. 3, *Early Modern Science.* Ed. by Katharine Park and Lorraine Daston, 320–40. Cambridge: Cambridge University Press.

Jones, Caroline A., and Peter Galison, eds. 1998. *Picturing Science, Producing Art.* New York: Routledge.

Kaiser, David. 2005. *Drawing Theories Apart: The Dispersion of Feynman Diagrams in Postwar Physics.* Chicago: University of Chicago Press.

Knight, David M. 1977. *Zoological Illustration: An Essay Towards a History of Printed Zoological Pictures.* Folkstone, England: Dawson.

Kohlstedt, Sally Gregory. 2010. *Teaching Children Science: Hands-on Nature Study in North America, 1890–1930.* Chicago: University of Chicago Press.

Kronick, David A. 1976. *A History of Scientific and Technical Periodicals: The Origins and Development of the Scientific and Technical Press, 1665–1790.* 2nd ed. Metuchen, N.J.: Scarecrow.

———. 1991. *Scientific and Technical Periodicals of the Seventeenth and Eighteenth Centuries: A Guide.* Metuchen, N.J.: Scarecrow.

———. 2004. *"Devant le deluge" and Other Essays on Early Modern Scientific Communication.* Lanham, Md.: Scarecrow.

Kusukawa, Sachiko. 1995. *The Transformation of Natural Philosophy: The Case of Philip Melanchthon.* Cambridge: Cambridge University Press.

———. 2000. "The *Historia Piscium* (1686)." *Notes and Records of the Royal Society of London* 54 (2): 179–97.

———. 2012. *Picturing the Book of Nature: Image, Text, and Argument in Sixteenth-Century Human Anatomy and Medical Botany.* Chicago: University of Chicago Press.

Kusukawa, Sachiko, and Ian Maclean, eds. 2006. *Transmitting Knowledge: Words, Images, and Instruments in Early Modern Europe.* Oxford: Oxford University Press.

Lackner, Michael, Iwo Amelung, and Joachim Kurtz, eds. 2001. *New Terms for New Ideas: Western Knowledge and Lexical Change in Late Imperial China.* Leiden: Brill.

Lackner, Michael, and Natascha Vittinghoff, eds. 2004. *Mapping Meanings: The Field of New Learning in Late Qing China.* Leiden: Brill.

Latour, Bruno. 1987. *Science in Action: How to Follow Scientists and Engineers through Society.* Cambridge, Mass.: Harvard University Press.

———. 1990. "Drawing Things Together." In *Representation in Scientific Practice.* Ed. by Michael Lynch and Steve Woolgar, 19–68. Cambridge, Mass.: MIT Press.

Law, John. 1986. "On the Methods of Long-Distance Control: Vessels, Navigation, and the Portuguese Route to India." In *Power, Action, and Belief: A New Sociology of Knowledge.* Ed. by John Law, 234–63. London: Routledge & Kegan Paul.

Lefèvre, Wolfgang, Jürgen Renn, and Urs Schoepflin, eds. 2003. *The Power of Images in Early Modern Science*. Basel: Birkhäuser.

Lightman, Bernard V. 2007. *Victorian Popularizers of Science: Designing Nature for New Audiences*. Chicago: University of Chicago Press.

Livingstone, David N. 2005. "Science, Text and Space: Thoughts on the Geography of Reading." *Transactions of the Institute of British Geographers* 30 (4): 391–401.

Lowood, Henry E., and Robin E. Rider. 1994. "Literary Technology and Typographic Culture: The Instrument of Print in Early Modern Science." *Perspectives on Science* 2 (1): 1–37.

Mandelbrote, Giles. 2000. "Scientific Books and Their Owners: A Survey to c. 1720." In *Thornton and Tully's Scientific Books, Libraries, and Collectors: A Study of Bibliography and the Book Trade in Relation to the History of Science*. 4th ed. Ed. by Andrew Hunter, 333–66. Aldershot, Hants, England: Ashgate.

Margócsy, Dániel. 2010. "'Refer to Folio and Number': Encyclopedias, the Exchange of Curiosities, and Practices of Identification before Linnaeus." *Journal of the History of Ideas* 71 (1): 63–89.

Martin, Henri-Jean, et al. 2000. *La naissance du livre moderne (XIVe-XVIIe siècles): Mise en page et mise en texte du livre français*. [Paris]: Éd. du Cercle de la librairie.

Martin, Henri-Jean, Roger Chartier, and Jean-Pierre Vivet, eds. 1982–86. *Histoire de l'édition française*. 4 vols. [Paris]: Promodis.

Mazzolini, Renato G., ed. 1993. *Non-Verbal Communication in Science prior to 1900*. Firenze: Olschki.

McGann, Jerome J. 1983. *A Critique of Modern Textual Criticism*. Chicago: University of Chicago Press.

McKenzie, D. F. 1969. "Printers of the Mind: Some Notes on Bibliographical Theories and Printing-House Practices." *Studies in Bibliography* 22: 1–75.

———. 1981. "Typography and Meaning: The Case of William Congreve." *Buch und Buchhandel in Europa im achtzehnten Jahrhundert: Fünftes Wolfenbütteler Symposium vom 1. bis 3. November 1977*. Ed. by Giles Barber and Bernhard Fabian, 81–125. Hamburg: Hauswedell.

———. 1986. *Bibliography and the Sociology of Texts*. London: British Library.

McLuhan, Marshall. 1962. *The Gutenberg Galaxy: The Making of Typographic Man*. [Toronto]: University of Toronto Press.

McNally, Peter F., ed. 1987. *The Advent of Printing: Historians of Science Respond to Elizabeth Eisenstein's "The Printing Press as an Agent of Change."* Montreal: Graduate School of Library and Information Studies, McGill University.

Meadows, A. J., ed. 1980. *Development of Science Publishing in Europe*. Amsterdam: Elsevier Science Publishers.

Montgomery, Scott L. 2000. *Science in Translation: Movements of Knowledge through Cultures and Time*. Chicago: University of Chicago Press.

Mosley, Adam. 2007. "Introduction: Objects, Texts, and Images in the History of Science." *Studies in History and Philosophy of Science Part A* 38 (2): 289–302.

Myers, Robin, and Michael Harris, eds. 1998. *Medicine, Mortality, and the Book Trade*. New Castle, Del.: Oak Knoll.

Ogilvie, Brian W. 2006. *The Science of Describing: Natural History in Renaissance Europe*. Chicago: University of Chicago Press.

Ong, Walter J. 1958. *Ramus: Method and the Decay of Dialogue: From the Art of Discourse to the Art of Reason.* Cambridge, Mass.: Harvard University Press.

———. 1967. *The Presence of the Word: Some Prolegomena for Cultural and Religious History.* New Haven: Yale University Press.

———. 1982. *Orality and Literacy: The Technologizing of the Word.* London: Methuen.

Overmier, Judith. 2000. "Scientific Book Collectors and Collections, Public and Private, 1720 to Date." In *Thornton and Tully's Scientific Books, Libraries, and Collectors: A Study of Bibliography and the Book Trade in Relation to the History of Science.* 4th ed. Ed. by Andrew Hunter, 367–91. Aldershot, Hants, England: Ashgate.

Pang, Alex Soojung-Kim. 2002. *Empire and the Sun: Victorian Solar Eclipse Expeditions.* Stanford, Calif.: Stanford University Press.

Pardo Tomás, José. 1991. *Ciencia y censura: La inquisición española y los libros científicos en los siglos XVI y XVII.* Madrid: Consejo Superior de Investigaciones Científicas.

Park, Katharine, and Lorraine Daston, eds. 2006. *The Cambridge History of Science.* Vol. 3, *Early Modern Science.* Cambridge: Cambridge University Press.

Porter, Roy, ed. 2003. *The Cambridge History of Science.* Vol. 4, *Eighteenth-Century Science.* Cambridge: Cambridge University Press.

Remmert, Volker R. (2005) 2011. *Widmung, Welterklärung und Wissenschaftslegitimierung: Titelbilder und ihre Funktionen in der Wissenschaftlichen Revolution.* Wiesbaden: Harrassowitz in Kommission. *Picturing the Scientific Revolution: Title Engravings in Early Modern Scientific Publications.* Trans. by Ben Kern. Philadelphia: Saint Joseph's University Press.

Rider, Robin E. 1998. "Shaping Information: Mathematics, Computing, and Typography." In *Inscribing Science: Scientific Texts and the Materiality of Communication.* Ed. by Timothy Lenoir, 39–54, 374–78. Stanford: Stanford University Press.

Rosenberg, Charles E., ed. 2003. *Right Living: An Anglo-American Tradition of Self-Help Medicine and Hygiene.* Baltimore: Johns Hopkins University Press.

Rosenberg, Daniel, et al. 2003. "Early Modern Information Overload." *Journal of the History of Ideas* [Special issue] 64 (1): 1–72.

Rudwick, Martin J. S. 1976. "The Emergence of a Visual Language for Geological Science, 1760–1840." *History of Science* 14 (3): 149–95.

———. 1992. *Scenes from Deep Time: Early Pictorial Representations of the Prehistoric World.* Chicago: University of Chicago Press.

Rupke, Nicolaas. 1999. "A Geography of Enlightenment: The Critical Reception of Alexander von Humboldt's Mexico Work." In *Geography and Enlightenment.* Ed. by David N. Livingstone and Charles W. J. Withers, 319–39. Chicago: University of Chicago Press.

———. 2000. "Translation Studies in the History of Science: The Example of *Vestiges.*" *British Journal for the History of Science* 33 (2): 209–22.

Schulten, Susan. 2001. *The Geographical Imagination in America, 1880–1950.* Chicago: University of Chicago Press.

Secord, James A. 2000. *Victorian Sensation: The Extraordinary Publication, Reception, and Secret Authorship of "Vestiges of the Natural History of Creation."* Chicago: University of Chicago Press.

Shea, William R., ed. 2000. *Science and the Visual Image in the Enlightenment.* [Canton, Mass.]: Science History Publications.

Sheets-Pyenson, Susan. 1981. "A Measure of Success: The Publication of Natural History Journals in Early Victorian Britain." *Publishing History* 9: 21–36.

———. 1985. "Popular Scientific Periodicals in Paris and London: The Emergence of a Low Scientific Culture, 1820–1875." *Annals of Science* 42 (6): 549–72.

Smith, Pamela H. 2006. "Art, Science, and Visual Culture in Early Modern Europe." *Isis* 97 (1): 83–100.

Topham, Jonathan R. 1998. "Beyond the 'Common Context': The Production and Reading of the Bridgewater Treatises." *Isis* 89 (2): 233–62.

———. 2000. "Scientific Publishing and the Reading of Science in Nineteenth-Century Britain: A Historiographical Survey and Guide to Sources." *Studies in History and Philosophy of Science Part A* 31 (4): 559–612.

Topham, Jonathan R., et al. 2000. "BJHS Special Section: Book History and the Sciences." *British Journal for the History of Science* 33 (2): 155–222.

Vande Walle, Willy F., and Kazuhiko Kasaya, eds. 2001. *Dodonaeus in Japan: Translation and the Scientific Mind in the Tokugawa Period.* Leuven: Leuven University Press.

Winkler, Mary G., and Albert Van Helden. 1992. "Representing the Heavens: Galileo and Visual Astronomy." *Isis* 83 (2): 195–217.

———. 1993. "Johannes Hevelius and the Visual Language of Astronomy." In *Renaissance and Revolution: Humanists, Scholars, Craftsmen, and Natural Philosophers in Early Modern Europe.* Ed. by J. V. Field and Frank A. J. L. James, 97–116. Cambridge: Cambridge University Press.

Wright, David. 2000. *Translating Science: The Transmission of Western Chemistry into Late Imperial China, 1840–1900.* Leiden: Brill.

Yeo, Richard R. 2001. *Encyclopaedic Visions: Scientific Dictionaries and Enlightenment Culture.* Cambridge: Cambridge University Press.

Contributors

RIMA D. APPLE is professor emerita at the University of Wisconsin–Madison. She has published extensively in women's history, the history of medicine and nursing, and the history of nutrition. Among her books are *Perfect Motherhood: Science and Childrearing in America* (Rutgers University Press, 2006) and *Vitamania: Vitamins in American Culture* (Rutgers University Press, 1996), which received the 1998 Kremers Award from the American Institute of the History of Pharmacy. She is currently studying the history of rural public health nursing.

JENNIFER J. CONNOR is a professor in the Faculty of Medicine and Department of History at Memorial University of Newfoundland. Her publications include her book, *Guardians of Medical Knowledge: The Genesis of the Medical Library Association* (Scarecrow, 2000), and articles in *Victorian Periodicals Review, Book History, History of the Book in Canada, Canadian Bulletin of Medical History*, and *Papers of the Bibliographical Society of Canada*.

MEGHAN DOHERTY is the director and curator of the Doris Ulmann Galleries and an assistant professor of art history at Berea College in Berea, Kentucky. Her research focuses on the connections between art and science as seen in the visual culture of the early Royal Society of London, and her current book project, *Carving Knowledge*, features studies of primary visual and written materials related to Hooke's *Micrographia*, Francis Willughby's Ornithology, and the *Philosophical Transactions of the Royal Society*.

GREGORY J. DOWNEY is a professor in the School of Journalism and Mass Communication and the School of Library and Information Studies at the University of Wisconsin–Madison. He is the author of *Telegraph Messenger Boys: Labor, Technology, and Geography, 1850–1950* (Routledge, 2002) and *Closed Captioning: Subtitling, Stenography, and the Digital Convergence of Text with Television* (Johns Hopkins University Press, 2008).

FLORENCE C. HSIA is an associate professor of history of science and integrated liberal studies at the University of Wisconsin–Madison. Her research interests include science and print culture, genres of scientific writing, natural knowledge-making in cross-cultural contexts, scientific travel, and the history of archival practices. She is the author of *Sojourners in a Strange Land: Jesuits and Their Scientific Missions in Late Imperial China* (University of Chicago Press, 2009).

CHERYL KNOTT is an associate professor in the School of Information Resources and Library Science at the University of Arizona. Her recent research focuses on the print culture of the environmental movement in the United States.

SALLY GREGORY KOHLSTEDT teaches and directs the History of Science and Technology Program at the University of Minnesota. She works on the history of science in American culture, with particular attention to locations where science reaches the public through institutions like museums and schools and engages diverse participants, particularly women. Especially relevant to her essay is her book *Teaching Children Science: Hands-On Nature Study in North America, 1890–1930* (University of Chicago Press, 2010) and a forthcoming edited volume with David Kaiser entitled *The American Century: Perspectives on Science, Technology, and Medicine*.

BERTRUM H. MACDONALD is professor of information management at Dalhousie University. He was the director of the School of Information Management from 1995 to 2003 and associate dean (research) in the Faculty of Management from 2002 to 2007. With a background in science, history of science, and information studies, he pursues research that investigates the diffusion and use of scientific and technical information.

KATE MCDOWELL is assistant professor at the Graduate School of Library and Information Science at the University of Illinois, Urbana-Champaign where her courses include youth services librarianship, history of readers, and storytelling. She has published articles in *Children and Libraries*, *Book History*, *Libraries and the Cultural Record*, and *Library Quarterly*. Her article "Surveying the Field: The Research Model of Women in Librarianship, 1882–1898" won the biennial 2010 Donald G. Davis Award of the American Library Association's Library History Round Table.

LYNN K. NYHART is Vilas-Bablich-Kelch Distinguished Achievement Professor of the history of science at the University of Wisconsin–Madison and the author of *Modern Nature: The Rise of the Biological Perspective in Germany* (University of Chicago Press, 2009), which analyzes the prehistory of German ecology in popular and museum science of the late nineteenth and early twentieth centuries.

ROBIN E. RIDER is senior lecturer in the Department of History of Science and the curator of special collections in the General Library System of the University of Wisconsin–Madison. Her research interests include early modern science, the printing and publishing of science, and the history of mathematics.

JAMES A. SECORD is professor in the Department of History and Philosophy of Science at the University of Cambridge. He is the author of *Victorian Sensation: The Extraordinary Publication, Reception and Secret Authorship of Vestiges of the Natural History of Creation* (University of Chicago Press, 2000).

STEPHEN L. VAUGHN is professor in the School of Journalism and Mass Communication at the University of Wisconsin–Madison, specializing in history. Vaughn's research and teaching interests include the history and social impact of new media, history of journalism, censorship, propaganda, modern entertainment, and American culture; the uses and misuses of history; and U.S. intellectual and political history. In addition to having written numerous books and scholarly articles, he is the editor of the *Encyclopedia of American Journalism* (Routledge, 2008), containing more than four hundred articles on the fourth estate.

PRINT CULTURE HISTORY
IN MODERN AMERICA

Series Editors

James P. Danky

Christine Pawley

Adam R. Nelson

Science in Print: Essays on the History of Science and the Culture of Print
Edited by Rima D. Apple, Gregory J. Downey, and Stephen L. Vaughn

Libraries as Agencies of Culture
Edited by Thomas Augst and Wayne Wiegand

Purity in Print: Book Censorship in America from the Gilded Age to the Computer Age,
Second Edition
Paul S. Boyer

Religion and the Culture of Print in Modern America
Edited by Charles L. Cohen and Paul S. Boyer

Women in Print: Essays on the Print Culture of American Women
from the Nineteenth and Twentieth Centuries
Edited by James P. Danky and Wayne A. Wiegand

Bookwomen: Creating an Empire in Children's Book Publishing, 1919–1939
Jacalyn Eddy

Apostles of Culture: The Public Librarian and American Society, 1876–1920
Lora Dee Garrison

Education and the Culture of Print in Modern America
Edited by Adam R. Nelson and John L. Rudolph